AutoCAD® 2000

Assessment Exam Preparation Manual

Alan J. Kalameja

*Developed for Autodesk, Inc.
Learning and Training*

Press

Thomson Learning™

Africa • Australia • Canada • Denmark • Japan • Mexico • New Zealand
Philippines • Puerto Rico • Singapore • Spain • United Kingdom • United States

NOTICE TO THE READER

Trademarks

Autodesk, AutoCAD and the AutoCAD logo are registered trademarks of Autodesk, Inc. Thomson Learning uses "Autodesk Press" with permission from Autodesk, Inc. for certain purposes. Windows is a trademark of the Microsoft Corporation. All other product names are acknowledged as trademarks of their respective owners.

Autodesk Press Staff

Executive Director: Alar Elken
Executive Editor: Sandy Clark
Aquisitions Editor: Michael Kopf
Developmental Editor: John Fisher
Executive Marketing Manager: Maura Theriault
Executive Production Manager: Mary Ellen Black
Production Manager: Larry Main
Art and Design Coordinator: Mary Beth Vought
Marketing Coordinator: Paula Collins
Technology Project Manager: Tom Smith
Editorial Assistant: Jasmine Hartman

Cover illustration by Scott Keidong's Image Enterprises
AutoCAD® image ©2000 Autodesk.

For more information, contact
Autodesk Press
3 Columbia Circle, Box 15-015
Albany, New York USA 12212-15015;
or find us on the World Wide Web at http://www.autodeskpress.com

Library of Congress Cataloging-in-Publication Data
ISBN 0-7668-2086-6

Contents

Preface | vii
General Information | ix
Taking the Exams | x
Exam Question Format | xi
Drawings on CD | xiv
Exam Registration | xv
Acknowledgements | xvi
About the Author | xvii

Chapter 1

AutoCAD 2000 Level I Exam Categories and Objectives 1

Level I Exam Categories | 2
Level I Exam Objectives | 3

Chapter 2

AutoCAD 2000 Level I Pretest 7

Answers to Level I Pretest | 31

Chapter 3

Analyzing 2D Drawings

33

Using Inquiry Commands	34
Finding the Area of an Enclosed Shape	35
Finding the Area of an Closed Polyline or Circle	36
Finding the Area of a Surface by Subtraction	37
Using the DIST (Distance) Command	38
Measuring Angles Using the DIST Command	39
Using the ID (Identify) Command	40
Using the LIST Command	41
Using the BOUNDARY Command	43
When to use BOUNDARY or PEDIT	44
Using the Properties Window	45
Tutorial Exercise - Extrude.Dwg	52
Tutorial Exercise - C-Lever.Dwg	57

Chapter 4

AutoCAD 2000 Level I

Practice Test

73

Answers to Level I Practice Test	99

Chapter 5

AutoCAD 2000 Level I

Exit Exam

101

Answers to Level I Exit Exam	127

Chapter 6

AutoCAD 2000 Level II Exam
Categories and Objectives   129

Level II Exam Categories   130
Level II Exam Objectives   131

Chapter 7

AutoCAD 2000 Level II
Pretest   135

Answers to Level II Pretest   161

Chapter 8

AutoCAD 2000 Level II
Practice Test   163

Answers to Level II Practice Test   188

Chapter 9

AutoCAD 2000 Level II
Exit Exam   191

Answers to Level II Exit Exam   219

Chapter 10
AutoCAD LT 2000
Level I and Level II Exams
Categories and Objectives   221

Answers to LT 2000 Level I Sample Questions   234
Answers to LT 2000 Level II Sample Questions   234

Appendix A
Exam Taking Tips
and Tricks   235

Appendix B
AutoCAD Assessment Exams
FAQ   237

Preface

A letter from the AutoCAD Exam Board:

Since the evolution of Computer-Aided Design (CAD), companies have been striving to maintain the most productive operators possible. To achieve this, many companies developed in-house tests to measure the skill and knowledge level of an operator. Those tests were designed around the particular application present at the company. Individuals seeking to have their CAD skills assessed sought out technical colleges, community colleges, and universities for CAD instruction leading to some type of certificate. However, because of differences in curriculum and grading policies, these certificates were recognized locally but not regionally and definitely not nationally. Other individuals received CAD training at one of the many authorized Autodesk Training Centers for AutoCAD located throughout the world. However, since this type of training is performed over a three to five-day period, operators were not given the time to gain a certain productivity level.

Assessment exams were developed in November, 1999 in response to CAD operator and employer demands. These exams are designed to measure your skill level at using various Autodesk products, including AutoCAD Release 14, AutoCAD 2000 and AutoCAD LT 2000. Each product exam is further broken down into Level I and Level II assessment exams.

This preparation manual begins with a preface containing general information regarding the exams. Topics include exam question format and how to register for the exams.

Chapters 1 through 5 of this manual concentrate on preparing for the AutoCAD 2000 Level I Assessment Exam. A pretest, practice test, and exit exam have been designed to give you a feel for the pace of the exam and more experience in handling different test problems and questions. A chapter on the use of all Inquiry commands is included, followed by a series of tutorials taking you through step-by-step instructions toward the completion of a problem.

Chapters 6 through 9 concentrate on preparing you for the Level II Assessment Exam. As with the Level I exam, a pretest, practice test, and exit exam have been designed for the Level II Exam to give you a realistic feel for what the exam will be like.

Chapter 10 concentrates on categories and objectives that define the AutoCAD LT 2000 exams. A few sample questions specific to AutoCAD LT 2000 are presented at the end of this chapter.

As this manual consists only of practice tests and exit exams designed to identify problem areas you may have, it is not intended to be used as a textbook or training guide. Refer to the AutoCAD 2000 Reference Manual or one of the many AutoCAD textbooks on the market for addressing those problem areas identified by the sample tests. AutoCAD users have a tendency to rely on a limited number of AutoCAD commands and options. Since the AutoCAD Assessment Exams are comprehensive, the coverage of important AutoCAD features is meant to be thorough.

The members of the AutoCAD Exam Board wish you success in your goals for assessing your Level I and Level II AutoCAD skills in AutoCAD 2000, LT 2000, or Release 14. We hope that this process will enable you to become more productive and competitive in today's global economy.

The AutoCAD Exam Board:

Robert Anderson
Autodesk, Inc.
San Rafael, CA

Brian Glidden
CAD and GIS Consultant
Lancaster, CA

Dan Abbott
Southern Maine Technical College
South Portland, ME

Alan Kalameja
Trident Technical College
Charleston, SC

General Information

Assessment exams based on Autodesk products are available for Release 14, AutoCAD 2000, and AutoCAD LT 2000. All of the product exams are further broken down into a Level I and Level II exam. Each exam requires a working knowledge of the specific Autodesk product and tests one's ability to use the software to modify and construct drawings in an efficient manner.

Since the questions designed for all Assessment Exams are independent of operating systems, knowledge of operating systems are not required. An individual should be familiar with the current release of AutoCAD for Windows 95, Windows 98, or Windows NT 4.0.

Questions regarding the use of AutoLISP are set aside for the Level II exams in AutoCAD Release 14 and AutoCAD 2000. Questions of this type are few and of the most general nature. The AutoCAD Exam Board has developed exams based on fundamental and general uses of 2D AutoCAD and an introduction to such topics as Customization, which is also a Level II exam category. One should not expect to be asked questions that are considered trivial or obscure in nature.

A minimum test score has not been set by the Exam Board because of the nature of an Assessment exam; namely to determine an individual's strengths and areas to improve on. However, a score of 85% or better is an indication that you have mastered most categories and objectives of the exam.

Taking the Exams

This manual has been developed for the experienced user wishing to assess his or her skill in Level I and/or Level II exams using AutoCAD Release 14, AutoCAD 2000, and AutoCAD LT 2000. An individual taking the AutoCAD Level I exam should have completed a beginning 2D class in AutoCAD (R14, 2000, LT 2000) in addition to actually using the software in a production environment for a period of at least 300 hours.

An individual who is taking the AutoCAD Level II exam should have completed an advanced 2D class in AutoCAD (R14, 2000, LT 2000) in addition to actually using the software in a production environment for a period of at least 600 hours.

Typical drawings found in the exams include site plans, machine parts, floor plans, gaskets, roof plans, building elevations and sprockets.

Individuals who might desire to have their current skills in AutoCAD assessed include students, industrial designers, architectural/civil/mechanical/electrical technicians, trainers/educators, and end-users that use the software for general applications.

Exam Question Format

The Level I and Level II Autodesk product exams (R14, 2000, LT 2000) are delivered electronically. This means all questions will require the use of a computer to answer the questions. Three types of questions may be asked in either the Level I or Level II exams.

The first type of questions is called "Single-Answer Multiple-Choice." For this type of question, there is only one answer of the four provided that best addresses the given question. The following is an example of a single answer multiple choice question:

What command is used to refresh the drawing screen without performing a regeneration?
(A) REDRAW
(B) REFRESH
(C) REGEN
(D) REGENAUTO

The correct answer for the question above is "A" since the REDRAW command is the only option of the four distracters presented that performs a screen refresh without performing a regeneration.

The second type of question asked is in the form of a hot-spot, which requires the individual to answer the question by picking the area of an image representing an actual AutoCAD dialog box. Study the following question accompanied by the dialog box illustration below:

Using the illustration provided below, pick the area of the Drawing Units dialog box designed to change the number of decimal places from 4 to 2.

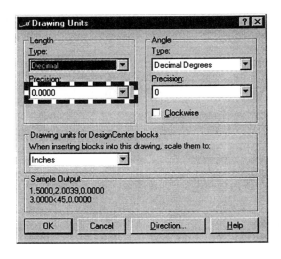

The individual would pick the area outlined in dashes above. The Precision area of the Drawing Units dialog box controls the number of decimal place units.

The third type of question asked is in the form of a graphical display accompanied by a single-answer multiple-choice question that requires the individual to choose the best answer from the list of four possible answers. Study the following question accompanied by the graphic illustration below:

What object snap mode would you use to construct a line from the corner at "A" to the corner at "B"?
  (A) Center
  (B) Corner
  (C) Endpoint
  (D) Nearest

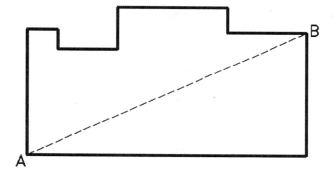

The correct answer is "C", Endpoint, which will construct line AB from the exact endpoints of the corners of the box in the illustration.

Drawings on CD

Most drawing questions require an individual to first load an existing drawing into AutoCAD 2000 or AutoCAD LT 2000. Directions on adding objects or editing the drawing follow. Questions are then asked which relate to each drawing. The following groups identify all drawings involved in the Level I and Level II pretests, practice tests, and exit exams in this manual:

Angle Block.Dwg	Floor-1.Dwg	Polygon.Dwg
Arc.Dwg	FMS.Dwg	Pool.Dwg
Base Plate.Dwg	Foundation.Dwg	Ratchet.Dwg
Block Walls.Dwg	Frame Guide.Dwg	Rotate.Dwg
Boundary.Dwg	Gasket.Dwg	Scale1.Dwg
Bracket Slide.Dwg	Geneva.Dwg	Scale2.Dwg
Bracket.Dwg	Gusset.Dwg	Scale3.Dwg
Building Plan 1.Dwg	Hitch.Dwg	Seal.Dwg
Building Plan 2.Dwg	House Plan.Dwg	Shape.Dwg
Building Plan 3.Dwg	Line.Dwg	Shield.Dwg
Buildings.Dwg	Locater.Dwg	Site.Dwg
Cam.Dwg	Metal Stamp.Dwg	Speaker.Dwg
Chain Links.Dwg	Mirror Pattern.Dwg	Special Cam.Dwg
Chamfer.Dwg	Multiple Chamfer.Dwg	Spline.Dwg
Color Pattern.Dwg	Office Plan.Dwg	Spring Mount.Dwg
Complex Expansion.Dwg	Pattern Template.Dwg	Stairs.Dwg
Coupling.Dwg	Perimeter.Dwg	Stamping.Dwg
Cover.Dwg	Plan.Dwg	Template.Dwg
Duplex.Dwg	Plat Plan.Dwg	View.Dwg
First Floor.Dwg	Plate.Dwg	

Create a folder called ASSESSMENT and copy all of the drawings above into it. Refer to this folder whenever an existing drawing file needs to be loaded in either the pretest, practice test, or exit exam of the Level I or Level II Assessment exams.

Exam Registration

Employers and applicants alike are seeking more guidance in measuring different levels of proficiency in various AutoCAD products. The complexity involved in electronic testing, coupled with the many and diverse applications for AutoCAD, pointed to the immediate need for a more sophisticated assessment program.

To register for the exam, use the following Web site:

www.autodesk.com/exams

Each exam costs $59.95 US dollars. When registering, you will be asked the following information: Your name, social security number, mailing address, phone number, organization you represent, and the type of exam you wish to register for; namely Level I or Level II exams based on either AutoCAD Release 14, AutoCAD 2000, or AutoCAD LT 2000.

Acknowledgements

Creating a specialized document such as this one has been no simple task, especially when individuals are looking to the assessment exams as a means of bettering themselves and gaining an edge for employment. As a result, many people were involved in the review process of this manual to insure its technical integrity and accuracy.

First, I would like to thank Kirsten Ludwig and Bob Anderson of Autodesk, Inc. for their guidance in the development of the Level I and Level II exam objectives for all product assessment exams.

From Delmar Publishers, Inc., I would like to thank the team of Sandy Clark, John Fisher, Mike Kopf and Mary Beth Vought for their continued support and guidance in putting this document together. I would also like to acknowledge Troy Reeves of the Delmar Southeast Marketing Region for his continued support and dedication in marketing this and other CAD related books to schools and end-users.

A complete and thorough job of technically editing the manuscript for this publication was performed by Dan Abbott of Southern Maine Technical College and a member of the Autodesk Exam Board, of which I will always be grateful. Thanks also to John Fisher who performed the copy edit of this document.

Finally, I would like to acknowledge the AutoCAD Exam Board of Robert Anderson, Dan Abbott, and Brian Glidden for their review of the Preface and Frequently Asked Questions.

About the Author

Alan J. Kalameja is the Department Head of Computer-Integrated Manufacturing (CIM) at Trident Technical College located in Charleston, South Carolina. The CIM Department at the College consists of Basic Construction Trades, CAD, Manufacturing, and Welding (which offers classes in robotic welding). He has been with the College for over 19 years and has been using AutoCAD since 1984 when Version 1.4 was available. He also directs the Authorized AutoCAD Training Center at Trident, which is charged with providing industry training to companies and firms at local and regional locations. Presently, he is an Autodesk Training Specialist in the Areas of AutoCAD, Mechanical Desktop, and Inventor. He has authored The AutoCAD Tutor for Engineering Graphics published by Delmar Publishers in the following software versions: Release 10/11; Release 12/13; Release 14; and AutoCAD 2000. A version of the AutoCAD Tutor for Engineering Graphics is also available through Delmar in AutoCAD LT 97 format. He has been a member of the AutoCAD Exam Board since November 1992.

This document was produced using PageMaker 6.5 by Adobe. All screen images were captured using Microsoft Photo Editor and saved in PCX format. Images of AutoCAD drawings were plotted out to a PLT format. Microsoft Word 97 was used to wordsmith the document before it was merged into PageMaker 6.5.

Notes

Chapter 1

AutoCAD 2000 Level I Exam Categories and Objectives

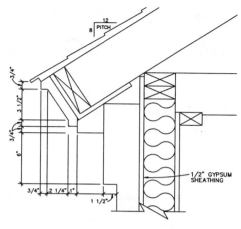

The AutoCAD 2000 Level I Assessment Exam consists of drawings and general knowledge questions that cover various AutoCAD topics. Inquiry commands are used to analyze each drawing question. This takes the form of using such commands as AREA, DIST, ID, and LIST for performing various calculations on each drawing. Knowledge of using the Properties dialog box and the ability to create selection sets using QSELECT command would also be helpful.

General knowledge questions may take the form of the following question types:

> Single answer multiple choice
> Hot spot

This chapter outlines the categories that make up the AutoCAD 2000 Level I Exam complete with the number of questions and a topic percentage that relates to the entire exam. Also, each category is further outlined with a detailed listing of the objectives an individual must master to be successful with the AutoCAD 2000 Level I Exam.

AutoCAD 2000
Level I Exam Categories

The AutoCAD 2000 Level I Exam consists of single answer multiple choice questions and hot spot questions. Use the chart below for a breakdown on the question categories, the number of questions per category, and the weight they carry in the AutoCAD 2000 Level I Exam.

AutoCAD 2000 Level I Exam Categories	Number of Questions	Percentage of Overall Score
1 Creating Drawing Template Files	2	4%
2 Display Commands	2	4%
3 Drawing Objects	12	24%
4 Extracting Drawing Information	1	2%
5 Editing	17	34%
6 Annotating Drawings	6	12%
7 Dimensioning a Drawing	3	6%
8 Managing Content	4	8%
9 Plotting	3	6%
Total	**50**	**100%**

AutoCAD 2000
Level I Exam Objectives

Category 1
Creating Drawing Template Files
Obj. 1.01   Create and delete layers
Obj. 1.02   Set layer properties

Category 2
Display Commands
Obj. 2.01   Use all options of the ZOOM command
Obj. 2.02   Save and restore a named view

Category 3
Drawings Objects
Obj. 3.01   Draw objects using Absolute Coordinates
Obj. 3.02   Draw objects using Relative Coordinates
Obj. 3.03   Draw objects using Polar Coordinates
Obj. 3.04   Use Direct Distance entry
Obj. 3.05   Use Polar Tracking
Obj. 3.06   Change Polar Tracking settings
Obj. 3.07   Draw a circle using all options
Obj. 3.08   Draw an arc using all options
Obj. 3.09   Set Point Size and Appearance
Obj. 3.10   Use the MEASURE and DIVIDE commands
Obj. 3.11   Use and identify Object Snaps
Obj. 3.12   **Acquire a point using Object Snap Tracking***

*Note: Objectives in bold apply only to the AutoCAD 2000 exam and NOT the
          AutoCAD LT 2000 exam

Category 4
Extracting Drawing Information
Obj. 4.01   Add and subtract areas using the AREA command

Category 5
Editing
Obj. 5.01   Select objects by Fence
Obj. 5.02   Use Previous and Last options to select objects
Obj. 5.03   Use Object Cycling
Obj. 5.04   Copy and move objects
Obj. 5.05   Use the OFFSET command
Obj. 5.06   Use the ARRAY command
Obj. 5.07   Use the MIRROR command
Obj. 5.08   Use the ROTATE command
Obj. 5.09   Use the SCALE command
Obj. 5.10   Use the STRETCH command
Obj. 5.11   Use the EXTEND command
Obj. 5.12   Use the TRIM command
Obj. 5.13   Use the FILLET and CHAMFER commands
Obj. 5.14   Use the BREAK command
Obj. 5.15   Use Grips to edit objects
Obj. 5.16   Use PEDIT to convert and join objects into a pline
Obj. 5.17   Use the Properties window

Category 6
Annotating Drawings
Obj. 6.01   Use the STYLE command to set and edit a text style
Obj. 6.02   Create single line and multiline text
Obj. 6.03   Format text using MTEXT
Obj. 6.04   Use the spell checker
Obj. 6.05   Use the BHATCH command
Obj. 6.06   Edit a hatch pattern

*Note: Objectives in bold apply only to the AutoCAD 2000 exam and NOT the
       AutoCAD LT 2000 exam

Category 7
Dimensioning a Drawing
Obj. 7.01	**Use QDIM** and DIMLINEAR
Obj. 7.02	Use the QLEADER command and its options
Obj. 7.03	Use the Dimension Style Manager

Category 8
Managing Content
Obj. 8.01	Create, insert and redefine a block
Obj. 8.02	Use MDE to copy objects and properties between drawings
Obj. 8.03	View and copy content with DesignCenter
Obj. 8.04	Use the PURGE command to reduce drawing size

Category 9
Plotting
Obj. 9.01	Setup a plot
Obj. 9.02	Use plot scale in model space and layouts
Obj. 9.03	Rename a layout

*Note: Objectives in bold apply only to the AutoCAD 2000 exam and NOT the AutoCAD LT 2000 exam

Notes

Chapter 2

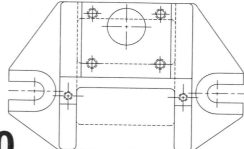

AutoCAD 2000
Level I
Pretest

This pretest is designed to assess your current AutoCAD 2000 skill and knowledge levels. You are presented with 50 questions. The AutoCAD 2000 Level I Pretest consists of general knowledge questions mixed in with performance-based drawing questions. Question types include single-answer multiple-choice and hot-spot areas. Numerous questions have been designed around actual images an individual would be confronted with in the production drawing environment. There is no time limit to complete this assessment test. However, this test is designed to be completed in 2 hours or less, which would demonstrate use of the software in a productive manner.

Two types of drawings are present in this pretest. Most of the drawings are already created up to a certain point. For these cases, open the drawing file and follow the steps that direct you to perform certain operations before attempting any of the questions that relate to the drawing. All drawings required for the test are provided on the CD supplied with this manual. Create a folder called \ASSESSMENT and load all drawing files there. Another type of drawing requires you to construct a new object from either the image provided or from written directions and answer the questions that follow the drawing to test your accuracy.

Work through this pretest at a good pace paying strict attention to the amount of time spent on each question. Answers for each Level I Pretest question are located at the end of this chapter.

Notes

Provide the best answer for each of the following questions.

1. Using the image of the Insert dialog box above, click on the area to break the block "CHAIR" up into individual objects.
Level I Objective 8.01

2. You open an existing drawing file. Grips are enabled. By default, what happens when an object is selected when no command is active?
   (A) The object highlights.
   (B) Blue outlined squares are displayed at key points along the object.
   (C) The object highlights along with blue outlined squares being displayed at key points along the object.
   (D) A single blue outlined square appears where the object was originally selected.
Level I Objective 5.15

3. You have magnified a portion of a drawing using ZOOM-Window. What scale factor is entered to perform a zoom at 1/10th of this current screen size?
   (A) 0.01X
   (B) 0.10X
   (C) 0.10
   (D) 0.01
Level I Objective 2.01

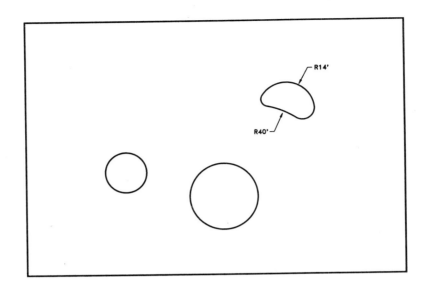

4. Open the existing drawing called POOL. Using the illustration above as a guide, construct two curves tangent to the two smaller circles; construct one curve with a radius of 40' and the other 14'. Offset the outline of the pool a distance of 1'-0". Pick a point anywhere outside of the pool as the side to offset. What is the area of the thin strip closest to?

    (A) 71 sq. ft.
    (B) 73 sq. ft.
    (C) 75 sq. ft.
    (D) 77 sq. ft.

Level I Objective 5.05

5. You need to create a number of layers in the current drawing file. From the following list, what name is NOT considered a valid layer name?

    (A) 1
    (B) $MECHANICAL
    (C) WIREFRAME@1
    (D) FIRST.FLOOR.PLAN

Level I Objective 1.01

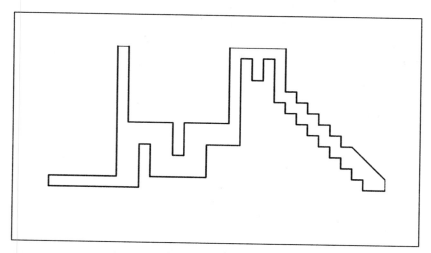

6. Open the drawing called FRAME GUIDE. Apply a fillet radius of 0.09 to all corners of the object. When finished, what is the total area of the Frame Guide?
    (A) 5.6502
    (B) 5.6507
    (C) 5.6512
    (D) 5.6517
Level I Objective 5.16

7. A drawing is constructed in real world units, (1 to 1). You want to plot the drawing at a scale of 1/4 its original size from Model Space. What value could be used for the Plot Scale?
    (A) 1=1
    (B) 0.25=1
    (C) 1=0.25
    (D) 0.25
Level I Objective 9.02

8. You construct a circle using the CIRCLE command along with the 2P option. What does this option prompt you for?
    (A) The radius of the circle.
    (B) The perimeter of the circle.
    (C) The circumference of the circle.
    (D) The endpoints of the circle's diameter.
Level I Objective 3.07

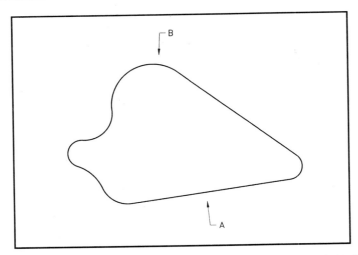

9. Open the existing drawing called SHAPE. Divide the shape into 27 equal parts.
   What is the distance from the Point labeled "A" to the Point labeled "B"?
   - (A) 10.07
   - (B) 10.12
   - (C) 10.17
   - (D) 10.22

Level I Objective 3.10

10. You have arranged two drawings using the Multiple Drawing Environment.
    What operations can be performed between these two drawings?
    - (A) copying, rotating, breaking
    - (B) copying, erasing, mirroring
    - (C) copying, moving, match properties
    - (D) copying, extending, match properties

Level I Objective 8.02

11. In a busy drawing, you construct a circle. You now want to mirror this circle.
    What option would you use at the "Select objects" prompt to select just the
    circle?
    - (A) All
    - (B) Last
    - (C) Object
    - (D) Previous

Level I Objective 5.02

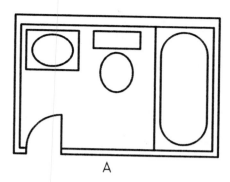
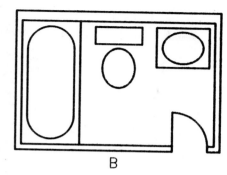

A   B

12. You construct a plan layout of the bathroom illustrated above in Figure A.
    However you need to modify the plan as shown in Figure B. What command
    is used to flip the image in the illustration at "A" to appear like the image in "B"?
    (A) COPY
    (B) FLIP
    (C) MIRROR
    (D) MOVE
Level I Objective 5.07

13. Click on the proper button in the illustration above designed to launch the
    AutoCAD DesignCenter.
Level I Objective 8.03

14. To cycle through objects for selection, which key do you hold down at the
    "Select objects:" prompt to activate Object Cycling?
    (A) ALT
    (B) CTRL
    (C) ESC
    (D) SHIFT
Level I Objective 5.03

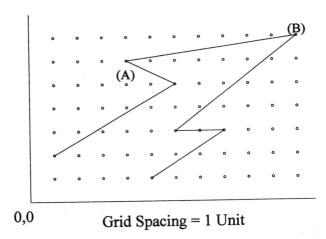

0,0   Grid Spacing = 1 Unit

15. In the figure above, what are the relative coordinates from Point "B" to Point "A"?

    (A) @-7.00,1.00
    (B) @1.00,7.00
    (C) @-7.00,-1.00
    (D) @7.00,-1.00

Level I Objective 3.02

16. Which statement is correct regarding the PURGE command?

    (A) It will delete a layer and all objects located on that layer.
    (B) It must be the first command used after entering the drawing editor.
    (C) It may be used to delete or erase blocks currently being displayed on a drawing.
    (D) It allows the user to delete unused items such as blocks, layers, and linetypes from the drawing database at any time.

Level I Objective 8.04

17. Which function key turns Polar Tracking mode on or off?

    (A) F9
    (B) F10
    (C) F11
    (D) F12

Level I Objective 3.05

18. You need to create several lines of text in a drawing. Each line of text will be
    a separate object. What command would you use to accomplish this task?
    (A) IMPORT
    (B) MTEXT
    (C) TEXT
    (D) TXT

Level I Objective 6.02

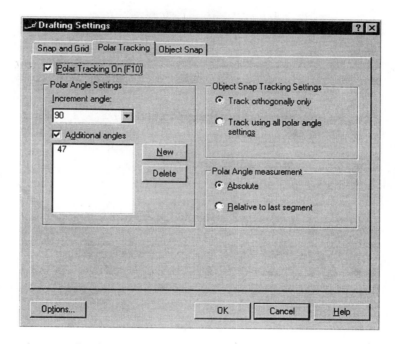

19. Click on the area in the Polar Tracking tab designed to set a standard angle.
    This will also allow you to snap to various multiples of that angle.

Level I Objective 3.06

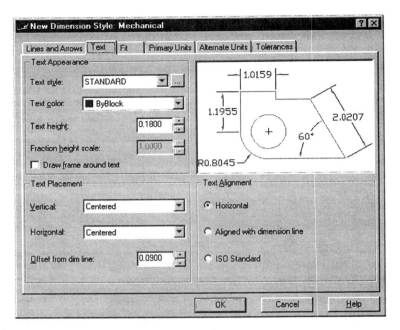

20. From the illustration above, click on the area of the Text tab of the New Dimension Style dialog box designed to place dimension text above the dimension line.
Level I Objective 7.03

21. What command is used to partially delete a portion of a line segment without the aid of the TRIM command?
     (A) BREAK
     (B) CUT
     (C) ERASE
     (D) SUBTRACT
Level I Objective 5.14

22. What command requires a boundary edge in order for one object to meet with another object?
     (A) COPY
     (B) EXTEND
     (C) LENGTHEN
     (D) TRIM
Level I Objective 5.11

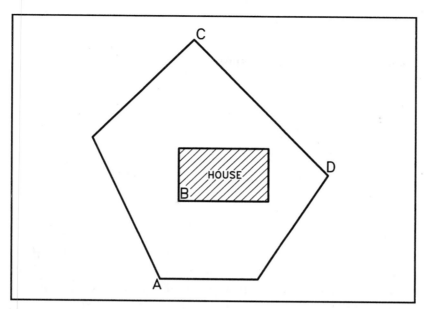

23. Open the existing drawing called SITE. Move the house 10'-0" to the left and 7'-0" up from the last point. What is the distance from the intersection at "B" to the intersection of vertex "D"?

    (A) 93'-4"
    (B) 93'-8"
    (C) 94'-0"
    (D) 94'-4"

Level I Objective 5.04

24. At the "Select objects:" prompt, you pick a number of points, creating a series of line segments that do not form a closed boundary. What Object Selection mode has just been described?

    (A) Crossing
    (B) Crossing Polygon
    (C) Fence
    (D) Window Polygon

Level I Objective 5.01

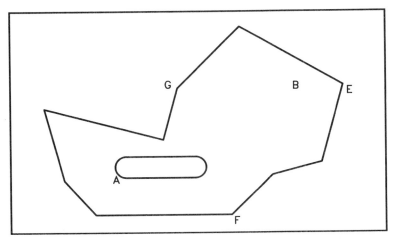

25. Open the existing drawing called ANGLE BLOCK. Turn on layers "Object" and "Dim". Move the slot 60 units to the right and 51 units up from the last point. What is the X-Y coordinate value of the center of arc "B"?

    (A) 176,109
    (B) 176,112
    (C) 179,112
    (D) 179,115

Level I Objective 1.02

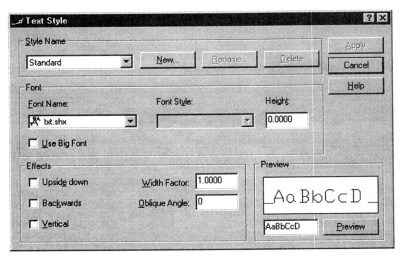

26. Click on the area of the Text Style dialog box designed to change to a different text style.

Level I Objective 6.01

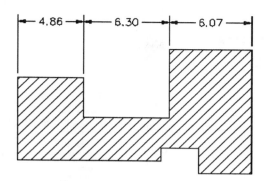

27. You activate the QDIM command. Dimensions need to be efficiently placed side-by-side as in the illustration above. What option of the QDIM command would you use to accomplish this task?

    (A) Baseline
    (B) Continuous
    (C) Ordinate
    (D) Staggered

Level I Objective 7.01

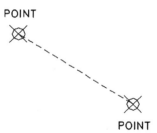

28. Two point objects are placed in the illustration above. A line segment needs to be constructed from one point to another. What Object Snap mode is used to perform this operation?

    (A) Center
    (B) Endpoint
    (C) Intersect
    (D) Node

Level I Objective 3.11

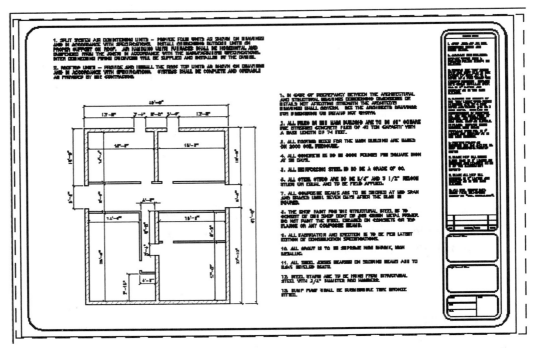

29. Open the drawing called BUILDING PLAN 1. After performing a search on all text, what is the name of the misspelled word?

    (A) EXKAVATION
    (B) HTE
    (C) STEL
    (D) VARRY

Level I Objective 6.04

30. You need to duplicate an object in a rectangular pattern. This requires you to enter the number of rows and columns. You also must enter the distance between rows and columns. What command allows you to perform this operation?

    (A) ARRAY
    (B) COPY
    (C) MIRROR
    (D) MOVE

Level I Objective 5.06

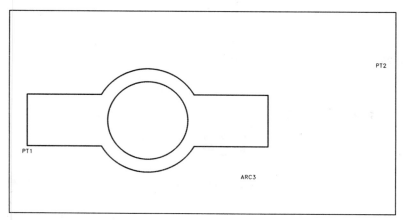

31. Open the drawing SCALE1. Increase the size of all yellow objects using a scale factor of 1.45. Use the endpoint at PT1 as the base point of the scale operation. When finished, what is the distance from the endpoint at PT2 to the center of ARC3?

    (A) 6.6838
    (B) 6.6842
    (C) 6.6846
    (D) 6.6850

Level I Objective 5.09

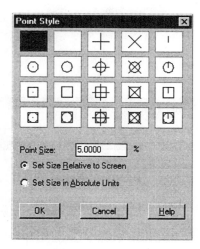

32. Click in the area of the Point Style dialog box designed to set the point display size as a percentage of the screen size.

Level I Objective 3.09

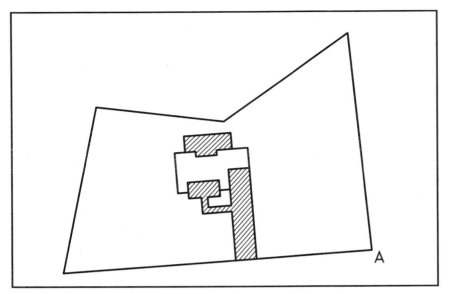

33. Open the existing drawing called PLAT PLAN. Stretch the intersection at vertex "A" a distance of 30'-0" directly to the right. What is the measurement of new angle at vertex "A"?

    (A) 78 degrees
    (B) 80 degrees
    (C) 82 degrees
    (D) 84 degrees

Level I Objective 5.10

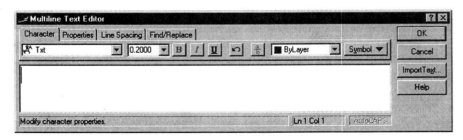

34. Click on the tab of the Multiline Text Editor dialog box designed take you to the area to use an existing text style.

Level I Objective 6.03

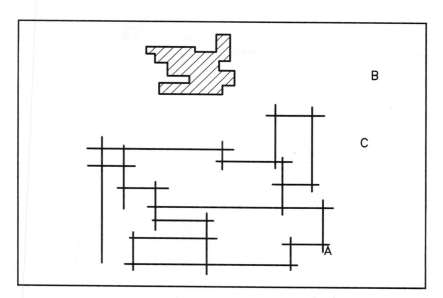

35. Open the existing drawing called TEMPLATE. Rotate all lines 270 degrees in the counterclockwise direction using the intersection at "A" as the base point. What is the X-Y coordinate value of intersection "B"?

    (A) 13.010,8.138
    (B) 13.014,8.142
    (C) 13.018,8.142
    (D) 13.018,8.146

Level I Objective 5.08

36. You enter the TRIM command. When prompted to select cutting edges at the "Select objects:" prompt, you press the ENTER key. What is the result?

    (A) It exits the TRIM command.
    (B) It produces the statement "Invalid."
    (C) The "Select objects:" prompt is displayed again.
    (D) All objects visible on the screen are selected as cutting edges.

Level I Objective 5.12

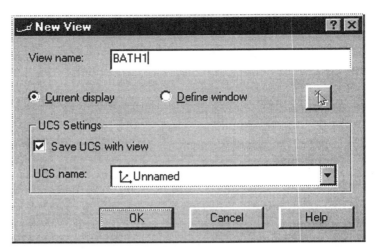

37. You open up a drawing. The total drawing is displayed. Click on the area of
    the New View dialog box designed to create a new view called "BATH1" from
    a designated area of the display screen.
Level I Objective 2.02

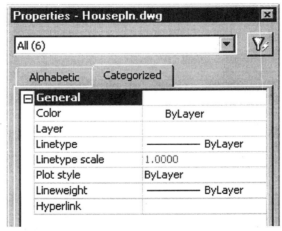

38. In a drawing, you select a block, a line, a circle, a line of text, and a number of
    arc objects. You then activate the Properties window. Click in the proper area
    of the Properties window above designed to display the current properties of
    just the line object.
Level I Objective 5.17

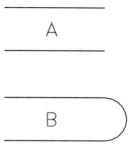

39. You click on the ends of the two parallel lines illustrated above in "A" and the ends are capped with a semi-circle in "B". What is the most efficient method of performing this operation?

    (A) using the FILLET command

    (B) using the 2P option of the CIRCLE command along with TRIM

    (C) using the TTR option of the CIRCLE command along with TRIM

    (D) using the S,C,R mode of the ARC command

Level I Objective 5.13

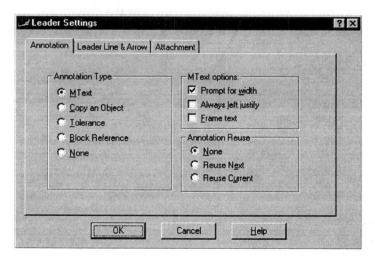

40. Click on the area of the Leader Settings dialog box designed to take you to the area that controls the placement of text alongside the leader.

Level I Objective 7.02

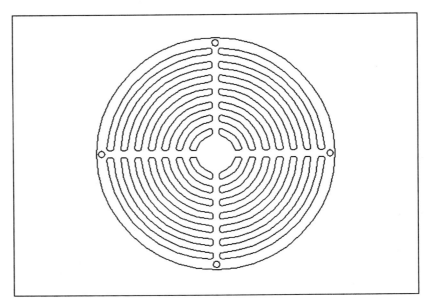

41.Open the existing drawing called COVER. All slots and holes inside the
perimeter of the large circle are made up of individual objects. Using the most
efficient means possible, calculate the area of the Cover with all slots and the
four holes removed. When finished, what is the calculated area of the cover?

(A) 35.1153
(B) 35.1159
(C) 35.1165
(D) 35.1171

Level I Objective 4.01

42. A line needs to be drawn at a distance of sixteen feet, four and five eighths inches
in the direction directly above the last known point. From the following list,
what coordinate entry is used to accomplish this task?

(A) @-16'4-5/8<90
(B) @16'-4-5/8<270
(C) @16'-4 5/8<270
(D) @16'4-5/8<90

Level I Objective 3.03

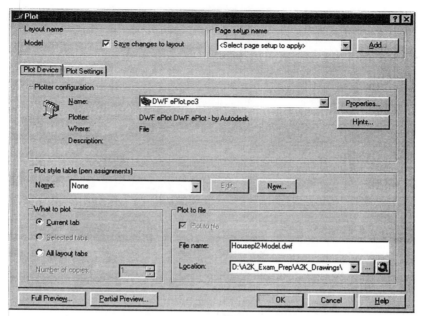

43. From the illustration above, click on the area of the Plot dialog box designed to change to a different type of plotting device.
Level I Objective 9.01

44. An object containing crosshatching is lengthened using the STRETCH command. What are the results of performing this stretch operation on an associative hatch pattern?

     (A) The hatch pattern stretches along with the object.

     (B) The hatch pattern explodes back into individual objects.

     (C) The hatch pattern stretches but an extra 90 degrees are added to the angle of the hatch pattern.

     (D) Nothing. The object stretches but the associative hatch pattern remains the same.

Level I Objective 6.06

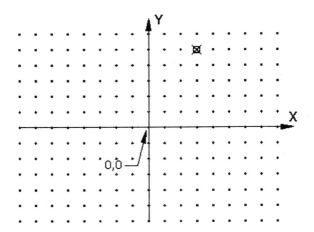

45. The illustration above of a Cartesian Coordinate grid. Each grid dot represents 1 unit. What is the X,Y coordinate value of the designated marker in the illustration above?

    (A) 5,3
    (B) 3,5
    (C) 6,10
    (D) 10,6

Level I Objective 3.01

46. You need to construct a line 10 units in length and in the 180 degree direction. You turn Ortho on, enter the LINE command and pick a point at the "Specify first point:" prompt. At the "Specify next point or [Undo]:" prompt, you move your cursor to the left and type "10." What mode of entry does this action describe?

    (A) Absolute Coordinate
    (B) Direct Distance
    (C) Polar Coordinate
    (D) Polar Tracking

Level I Objective 3.04

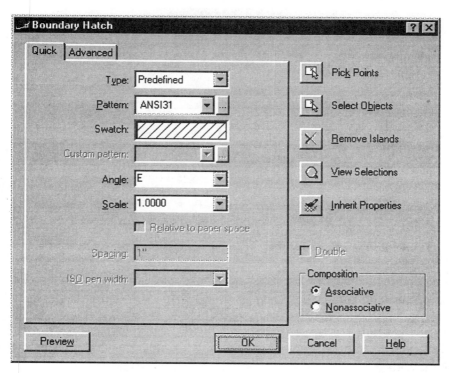

47. After activating the Boundary Hatch dialog box, you pick a hatch pattern, identify an internal point in a drawing, and preview the results. Click on the area of the Boundary Hatch dialog box designed to add the hatch pattern to the drawing.

Level I Objective 6.05

48. You have constructed a polyline with the RECTANG command. You need to acquire a point located on the upper right corner of the rectangle for the purpose of using Object Snap Tracking. How is this point acquired?
    - (A) moving your cursor directly over the point and pausing
    - (B) moving your cursor directly over and selecting the point
    - (C) moving your cursor directly over the point, pressing the CTRL key, and selecting the point
    - (D) moving your cursor directly over the point, pressing the SHIFT key, and selecting the point

Level I Objective 3.12

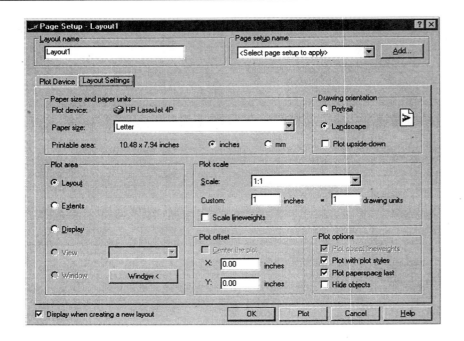

49. Click on the area of the Page Setup dialog box designed to change the name of
    the layout.
Level I Objective 9.03

50. Start a new drawing. Construct an arc with the following parameters:
    Radius = 6.00
    Start = 4.00,4.00
    End = 11.00,2.50
    With the arc created, what is the total length of the arc?
    (A) 7.6707
    (B) 7.6712
    (C) 7.6717
    (D) 7.6722
Level I Objective 3.08

Answers to the AutoCAD 2000 Level I Pretest

1. Place a check in the Explode box.

2. C

3. B

4. B

5. D

6. B

7. B

8. D

9. C

10. C

11. B

12. C

13. Click on this button:

14. B

15. C

16. D

17. B

18. C

19. Click on the arrow underneath Increment Angle:

20. Under the Text Placement area, click in the box next to Vertical and change the value to Above.

21. A

22. B

23. D

24. C

25. D

26. Click in the box underneath Style Name:

27. B

28. D

29. C

30. A

31. B

32. Click on the radio button adjacent to Set Size Relative to Screen.

33. B

34. Click on the Properties tab of the Multiline Text Editor dialog box.

35. D

36. D

37. Click on the Define Window radio button.

38. Click in the top box labeled "all(6)" and then select "Line".

39. A

40. Click on the Attachment tab of the Leader Settings dialog box.

41. C

42. D

43. In the Plotter configuration area of the Plot dialog box, click in the box adjacent to Name.

44. A

45. B

46. B

47. Click on the OK button.

48. A

49. Click in the box underneath Layout name.

50. A

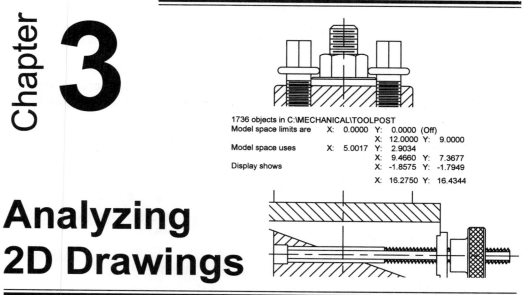

1736 objects in C:\MECHANICAL\TOOLPOST
Model space limits are   X: 0.0000 Y: 0.0000 (Off)
                           X: 12.0000 Y: 9.0000
Model space uses   X: 5.0017 Y: 2.9034
                           X: 9.4660 Y: 7.3677
Display shows   X: -1.8575 Y: -1.7949
                           X: 16.2750 Y: 16.4344

Analyzing
2D Drawings

Completed drawings are usually plotted out and checked with scales for accuracy. A proper computer-aided design system is equipped with a series of commands to calculate distances and angles of selected objects. Surface areas may be calculated for complex geometric shapes.

The next series of pages highlight all Inquiry commands and how they are used to display useful information about an object or group of objects. The Properties dialog box is also explained in great detail.

Use the information in this chapter to become more familiar with all Inquiry commands (especially AREA, DIST, ID and LIST) and the Properties window.

Using Inquiry Commands

AutoCAD's Inquiry commands may be selected from the Inquiry Toolbar, may be keyed in at the keyboard, or may be selected from the pulldown menu bar or side bar screen menus as illustrated below. The following is a listing of the Inquiry commands with a short description of each:

AREA - used to calculate the surface area from a series of points or by selecting a polyline or circle. Multiple entities may be added or subtracted to calculate the area of an object with holes and cutouts.

DIST - calculates the distance between two points. Also provides the delta X,Y,Z coordinate values, the angle in the X-Y plane, and the angle from the X-Y plane.

HELP - provides online help for any command. May be entered at the keyboard or selected from a dialog box.

ID - displays the X,Y,Z absolute coordinate of a selected point.

LIST - displays key information depending on the entity selected.

STATUS - displays important information on the current drawing.

TIME - displays the time spent in the drawing editor.

The next series of pages gives a detailed description on how these commands are used with different AutoCAD objects.

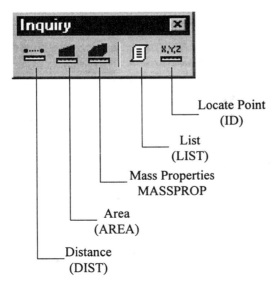

Locate Point
(ID)

List
(LIST)

Mass Properties
MASSPROP

Area
(AREA)

Distance
(DIST)

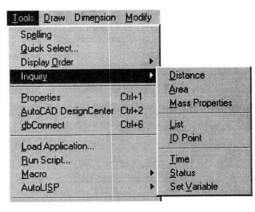

Finding the Area of an Enclosed Shape

The AREA command is used to calculate the area through the selection of a series of points. Select the endpoints of all vertices of the image illustrated below (OSNAP-Endpoint by default is active). Once the first point is selected along with the remaining points in either a clockwise or counterclockwise pattern, the command prompt "Next corner point..." is followed by the ENTER key to calculate the area of the shape (Displayed below). Along with the area is a calculation for the perimeter. Use the illustrations below to gain a better understanding of the prompt sequence used for finding the area by identifying a series of points.

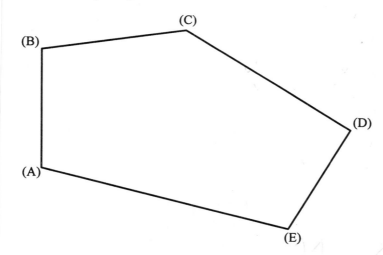

Command: **AA** *(For AREA)*
Specify first corner point or [Object/Add/Subtract]: *(Pick the endpoint at "A")*
Specify next corner point or press ENTER for total: *(Pick the endpoint at "B")*
Specify next corner point or press ENTER for total: *(Pick the endpoint at "C")*
Specify next corner point or press ENTER for total: *(Pick the endpoint at "D")*
Specify next corner point or press ENTER for total: *(Pick the endpoint at "E")*
Specify next corner point or press ENTER for total: *(Pick the endpoint at "A")*
Specify next corner point or press ENTER for total: *(Press ENTER to calculate the area)*

Area = 25.25, Perimeter = 20.35

Finding the Area of an Enclosed Polyline or a Circle

(A)

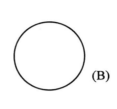
(B)

In the previous example, the area of an enclosed shape was found by using the AREA command and identifying the corners or intersections of the enclosed area by a series of points. For a complex area, this could be a very tedious operation. As a result, the AREA command has a built-in Object option that will calculate the area and perimeter of a polyline and the area and circumference of a circle. Study the illustrations above and below for these operations.

Finding the area of a polyline can only be accomplished if one of the following conditions are satisfied:

- The shape must have already been constructed using the PLINE command.

- The shape must have already been converted into a polyline using the Pedit command if originally constructed out of individual objects.

- The BOUNDARY command can often be used to quickly create a closed polyline.

Command: **AA** *(For AREA)*
Specify first corner point or [Object/Add/Subtract]: **O** *(For Object)*
Select objects: *(Select the polyline at "A")*
Area = 24.88, Perimeter = 19.51

Command: **AA** *(For AREA)*
Specify first corner point or [Object/Add/Subtract]: **O** *(For Object)*
Select objects: *(Select the circle at "B")*
Area = 7.07, Circumference = 9.42

Finding the Area of a Surface by Subtraction

The steps used to calculate the total surface area are: (1) calculate the area of the outline; and (2) subtract the objects inside of the outline. All individual objects, with the exception of circles, must first be converted into polylines using the PEDIT command. Next, the overall area is found and added to the database using the Add mode of the AREA command. Add mode is exited and the inner objects are removed using the Subtract mode of the AREA command. Remember, all objects must be in the form of a circle or polyline. This means the inner shape at "B" must also be converted into a polyline using the PEDIT command before calculating the area. Care must

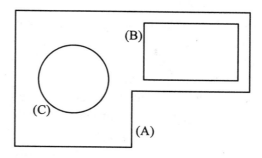

be taken when selecting the objects to subtract. If an object is selected twice, it is subtracted twice and may yield an inaccurate area in the final calculation.

For the image above, the total area with the circle and rectangle removed is 30.4314.

Command: **AA** *(For AREA)*
Specify first corner point or [Object/Add/Subtract]: **A** *(For Add)*
Specify first corner point or [Object/Subtract]: **O** *(For Object)*
(ADD mode) Select objects: *(Select the polyline at "A")*
Area = 47.5000, Perimeter = 32.0000
Total area = 47.5000
(ADD mode) Select objects: *(Press ENTER to exit Add mode)*
Specify first corner point or [Object/Subtract]: **S** *(For Subtract)*
Specify first corner point or [Object/Add]: **O** *(For Object)*
(SUBTRACT mode) Select objects: *(Select the polyline at "B")*
Area = 10.0000, Perimeter = 13.0000
Total area = 37.5000
(SUBTRACT mode) Select objects: *(Select the circle at "C")*
Area = 7.0686, Circumference = 9.4248
Total area = 30.4314
(SUBTRACT mode) Select objects:*(Press ENTER to exit Subtract mode)*
Specify first corner point or [Object/Add]: *(Press ENTER to exit)*

Using the DIST (Distance) Command

The DIST command calculates the linear distance between two points on an object whether it be the length of a line, the distance between two points, or the distance from the quadrant of one circle to the quadrant of another circle. The following information is also supplied when using the Dist command: the angle in the X-Y plane; the angle from the X-Y plane; the delta X, Y, and Z coordinate values. The angle in the X-Y plane is given in the current angular mode set by the Units command. The delta X, Y, and Z coordinate is a relative coordinate value taken from the first point identified by the DIST command to the second point.

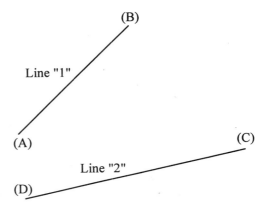

Command: **DI** *(For DIST)*
Specify first point: *(Pick the endpoint at "A")*
Specify second point: *(Pick the endpoint at "B")*
Distance = 6.36, Angle in XY Plane = 45.0000,
Angle from XY Plane = 0.0000
Delta X = 4.50, Delta Y = 4.50,   Delta Z = 0.00

Command: **DI** *(For DIST)*
Specify first point: *(Pick the endpoint at "C")*
Specify second point: *(Pick the endpoint at "D")*
Distance = 9.14, Angle in XY Plane = 192.7500,
Angle from XY Plane = 0.0000
Delta X = -8.91, Delta Y = -2.02,   Delta Z = 0.00

Measuring Angles Using the DIST Command

On the previous page, it was pointed out that the DIST command yields information regarding distance, delta X, Y coordinate values, and angle information. Of particular interest is the angle in the X-Y plane formed between two points. In the illustration at the right, picking the endpoint of the line segment at "A" as the first point followed by the endpoint of the line segment at "B" as the second point displays an angle of 42 degrees. This angle is

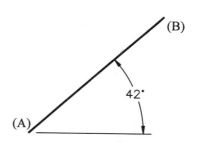

formed from an imaginary horizontal line drawn from the endpoint of the line segment at "A" in the zero direction.

Care needs to be taken when using the DIST command to find the angle of a line segment. If when using the DIST command, the endpoint of the line segment at "B" is selected as the first point followed by the endpoint of the segment at "A" for the second point, a new angle in the X-Y plane of 222 degrees is given. In the illustration at the right, the angle is calculated by constructing a horizontal line from the endpoint at "B," the new first point of the DIST command. This horizonal line is drawn in the zero direction. Notice the rela-

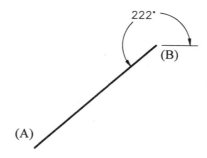

tionship of the line segment to the horizontal baseline. Be careful identifying the order when selecting line segment endpoints for extracting angular information.

Using the ID (Identify) Command

The ID command is one of the more straightforward of the Inquiry commands. ID stands for "Identify" and allows the user to obtain the current absolute coordinate listing of a point.

The coordinate value of the center of the circle at "A" was found by using ID and the OSNAP-Center mode; the coordinate value of the starting point of text string "B" was found using ID and the OSNAP-Insert mode; the coordinate value of the endpoint of line segment "C" was found using ID and the OSNAP-Endpoint mode; the coordinate value of the midpoint of line segment at "CD" was found by using ID

(A)

(B)
IDENTIFYING TEXT

(C)

(D)

(E)

and the OSNAP-Midpoint mode; and the coordinate value of the current position of point "E" was found by using ID and the OSNAP-Node mode.

Command: **ID**
Specify point: **Cen**
of *(Select the edge of circle "A")*
X = 2.00   Y = 7.00   Z = 0.00

Command: **ID**
Specify point: **Ins**
of *(Select the text at "B")*
X = 5.54   Y = 7.67   Z = 0.00

Command: **ID**
Specify point: **End**
of *(Select the line at "C)*
X = 8.63   Y = 4.83   Z = 0.00

Command: **ID**
Specify point: **Mid**
of *(Select line "CD")*
X = 5.13   Y = 3.08   Z = 0.00

Command: **ID**
Specify point: **Nod**
of *(Select the point at "E")*
X = 9.98   Y = 1.98   Z = 0.00

Using the LIST Command

Use the LIST command to obtain information about an object or group of objects. In the figure below, two rectangles are displayed along with a circle. Are the rectangles made up of individual line segments or a polyline object? Using the LIST command on each object informs you the first rectangle at "A" is a polyline, the circle lists as a circle, and the second rectangle is actually a block reference. In addition to the object type, you also can obtain key information such as the layer the object resides on, area and perimeter information for polylines; area and circumference information for circles. Study the prompt sequences on the next page for using the LIST command and observe the results displayed on each object type.

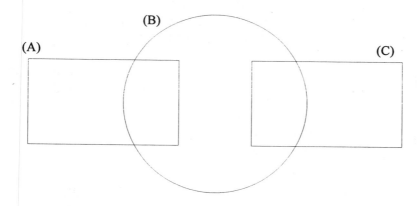

Command: **LI** *(For LIST)*
Select objects: *(Select the objects at "A", "B" and "C" in order)*
Select objects: *(Press ENTER to list the information on each object)*

LWPOLYLINE Layer: "0"
Space: Model space
Handle = 3A
Closed
Constant width   0.0000
area 3.3835
perimeter 7.6235
at point X=4.7002 Y=4.1846 Z=0.0000
at point X=7.1049 Y= 4.1846 Z=0.0000
at point X=7.1049 Y=5.5916 Z=0.0000
at point X=4.7002 Y=5.5916 Z=0.0000

CIRCLE   Layer: "0"
Space: Model space
Handle = 2B
center point, X=7.6879 Y=4.8881 Z=0.0000
radius   1.4719
circumference   9.2479
area   6.8058

BLOCK REFERENCE Layer: "0"
Space: Model space
Handle = 34
"rec"
at point, X=10.6756 Y=4.1846 Z=0.0000
X scale factor   -1.0000
Y scale factor   1.0000
rotation angle   0
Z scale factor   1.0000

Using the BOUNDARY Command

For creating polyline objects from existing geometry, BOUNDARY is the most efficient command. Choose this command from the Draw pull-down menu followed by Boundary... This action launches the Boundary Creation dialog box illustrated at the right. The Pick Points button will be used to create the boundary.

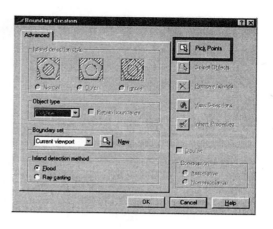

The object at the right consists of 88 individual objects. To convert each shape into a series of individual polyline objects, the PEDIT command would have to be used 6 times. Rather than use the PEDIT command, the BOUNDARY command allows you to select an internal point like Point "A" illustrated at the right. All objects will highlight after picking this point.

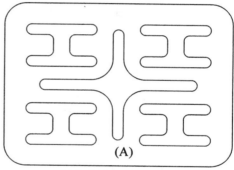

88 Individual Objects

With the objects highlighted, clicking the OK button of the Boundary Creation dialog box traces a polyline object over each shape. This results in the creation of 6 polylines in a single operation. Because BOUNDARY traces polylines by default over each shape in the current layer, it is considered good practice to place the polylines on a dedicated layer to separate the individual objects from the newly created polylines.

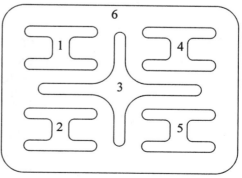

6 Polyline Objects

When to use **BOUNDARY** or **PEDIT**

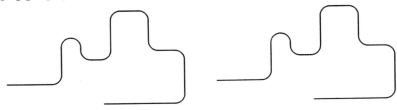

21 Individual Objects   1 Polyline Object after using the
                                                        PEDIT command

Illustrated above is an example of when you must use the PEDIT command instead of the BOUNDARY command to convert individual objects into a single polyline. The object at the left consists of 21 individual objects. The key is that the object does not form a closed shape. The BOUNDARY command requires that the individual objects close and for this reason, the PEDIT command must be used to covert the 21 individual objects into a single polyline.

Illustrated at the right is another example of determinining which command is better to use; PEDIT or BOUNDARY. The BOUNDARY command would be the command to use since the 12 individual objects do form a closed shape. The BOUNDARY command will ignore the overshoots and create a single polyline of the inner shape. The PEDIT command could not be used in this case since the lines do not form an exact corner, which is required when using PEDIT.

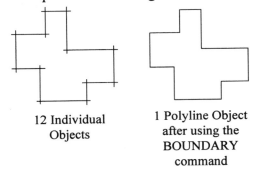

12 Individual   1 Polyline Object
Objects   after using the
                         BOUNDARY
                         command

The 12 objects at the right could not be coverted into a single continuous polyline object by using either the PEDIT or BOUNDARY commands. The presence of the gap would eliminate using the BOUNDARY command. The overshoots would not allow all lines to be connected together using the PEDIT command.

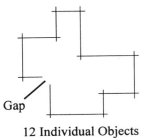

Gap

12 Individual Objects

Using the Properties Window

At times, objects are drawn on the wrong layer, color, or even in the wrong linetype. The length of line segments are incorrect, or the radius values of circles and arcs are incorrect. A series of tools is available to modify the properties of these objects eliminating the need to erase these objects and reconstruct them to their correct specifications. Illustrated below are three line segments. One of the line segments has been pre-selected, as shown by the highlighted appearance and the presence of grips.

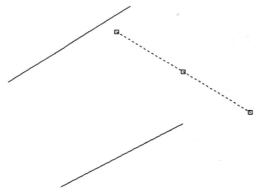

Clicking on the Properties button on the Standard toolbar below displays the Properties window at "A" on the next page. This window displays information about the object already selected; in this case the information is about the line segment, which is identified at the top of the dialog box. Two tabs are present in the dialog box. They allow you to easily view the information about the line. The first tab lists properties of the line alphabetically (see figure "A" on the next page).

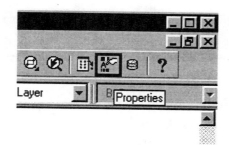

The second tab displays information in a series of categories, as in "B" below. Notice, in the Categorized tab, that two sets of information are provided about the selected object. One area deals with General properties to change. This area allows you to change the color, layer, linetype, linetype scale, etc. of the selected object. The second area gives information about the selected object. Since a single line was selected, the starting and ending X, Y, Z values are given. As modifications are made to these values, other information displayed automatically updates, such as length and angle of the line. You cannot make modifications to values that appear grayed out.

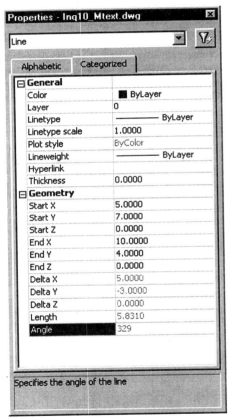

A   B

When changing a selected object to a different layer, click in the layer field illustrated on the next page in "C" and the current layer displays (in this example, Layer 0). Clicking the down arrow displays all layers defined in the drawing. Clicking on one of these layers changes the selected object to a different layer.

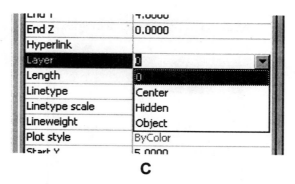

C

Illustrated below are the same three line segments. This time, all three segments have been pre-selected, as shown by the highlighted appearance and the presence of grips.

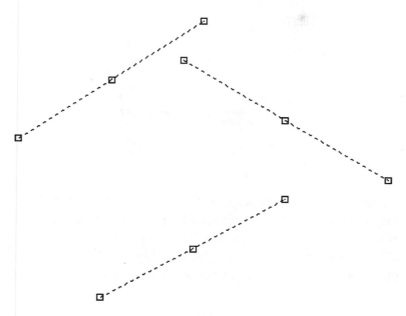

Clicking on the Properties button on the Standard toolbar displays the Properties window on the next page at "D". At the top of the dialog box, the three lines are identified. In the Alphabetic tab, you can change the color, layer, linetype, and other general properties of all three lines. However, you are unable to enter the Start and End X, Y, and Z values, and there is no length or angle information.

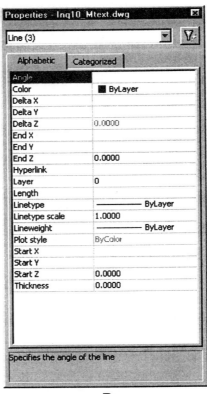

D

In the figure illustrated below, an arc, circle, and line are pre-selected along with the display of grips. Activating the Properties window on the next page at "E" displays a number of object types at the top of the dialog box. You can click which object or group of objects to modify. With "All (3)" highlighted, you can change the general properties, such as layer and linetype, but not any geometry settings.

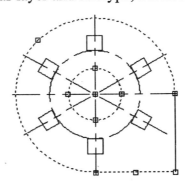

E

What if you need to increase the radius of the circle to 1.25 units? Click on the Circle object type at the top of the Properties window. The full complement of general and geometry settings is present for you to modify. Click in the Radius field and change the current value to 1.25 units. Pressing ENTER automatically updates the other geometry settings in addition to the actual object in the drawing (see the illustration below). When finished, dismiss this window to return to the drawing.

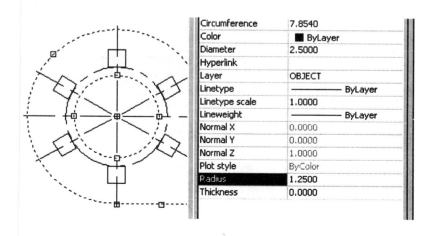

You can also open the Properties window by choosing Properties from the Modify pull-down menu illustrated below at "F". This displays the Properties window in "G". Notice that "No selection" is listed at the top of the dialog box, meaning that no objects have been selected to modify. Even though nothing is selected, you can still change the current color, linetype, and even make a layer current.

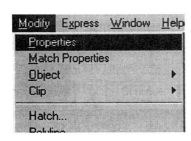

F

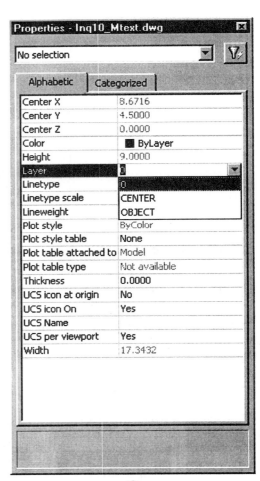

G

Clicking on the Quick Select button below displays the Quick Select dialog box. Use this dialog box to build a selection set to modify its object properties.

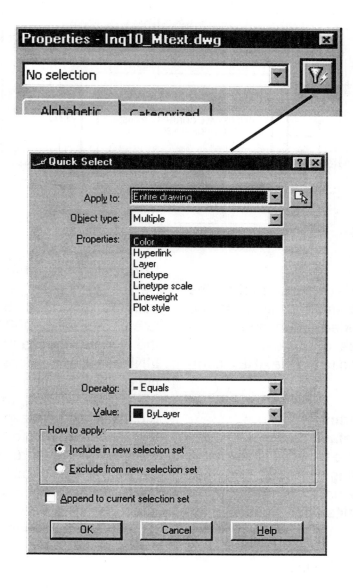

Tutorial Exercise
Extrude.Dwg

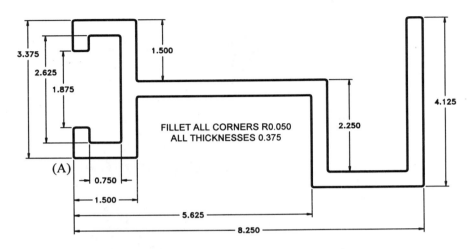

FILLET ALL CORNERS R0.050
ALL THICKNESSES 0.375

Purpose:
This tutorial is designed to show the user various methods in constructing the extruded pattern above. The surface area of the extrusion will also be found using the AREA command.

System Settings:
Keep the default drawing limit settings. Use the Drawing Units dialog box and change the number of decimal places past the zero from 4 units to 3 units. Use English units.

Layers:
Create the following layers:

Name	Color	Linetype
Boundary	Magenta	Continuous
Object	Yellow	Center

Suggested Commands:
Begin drawing the extrusion with point "A" illustrated above at absolute coordinate 2.000,3.000. Use the Direct Distance mode and the LINE command to construct the profile of the Extrusion. Then use the FILLET command to create the 0.050 radius rounds at all corners of the extrusion. Before calculating the area of the extrusion, create a single polyline through the use of the BOUNDRY command. This will allow the AREA command to be used in a more productive way.

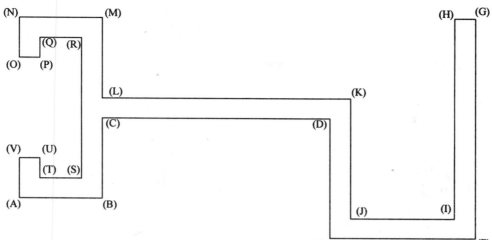

Step #1

First make the Object layer current. One method of constructing the extrusion above is to use the measurements on the previous page to calculate a series of polar coordinate distances. The Direct Distance mode could also be use to complete the profile.

Command: **L** *(For LINE)*
Specify first point: 2.000,3.000 *(Starting at "A")*
Specify next point or [Undo]: **@1.500<0** *(To "B")*
Specify next point or [Undo]: **@1.500<90** *(To "C")*
Specify next point or [Close/Undo]: **@4.125<0** *(To "D")*
Specify next point or [Close/Undo]: **@2.250<270** *(To "E")*
Specify next point or [Close/Undo]: **@2.625<0** *(To "F")*
Specify next point or [Close/Undo]: **@4.125<90** *(To "G")*
Specify next point or [Close/Undo]: **@0.375<180** *(To "H")*

Specify next point or [Close/Undo]: **@3.750<270** *(To "I")*
Specify next point or [Close/Undo]: **@1.875<180** *(To "J")*
Specify next point or [Close/Undo]: **@2.250<90** *(To "K")*
Specify next point or [Close/Undo]: **@4.500<180** *(To "L")*
Specify next point or [Close/Undo]: **@1.500<90** *(To "M")*
Specify next point or [Close/Undo]: **@1.500<180** *(To "N")*
Specify next point or [Close/Undo]: **@0.750<270** *(To "O")*
Specify next point or [Close/Undo]: **@0.375<0** *(To "P")*
Specify next point or [Close/Undo]: **@0.375<90** *(To "Q")*
Specify next point or [Close/Undo]: **@0.750<0** *(To "R")*
Specify next point or [Close/Undo]: **@2.625<270** *(To "S")*
Specify next point or [Close/Undo]: **@0.750<180** *(To "T")*
Specify next point or [Close/Undo]: **@0.375<90** *(To "U")*
Specify next point or [Close/Undo]: **@0.375<180** *(To "V")*
Specify next point or [Close/Undo]: **C** *(To Close)*

Step #2

Rather than converting individual objects into one polyline using the PEDIT command and Join option, you could use a more efficient means of creating a polyline: the BOUNDARY command. First make the Boundry layer current. Choosing Boundary... from the Draw pull-down menu at the right activates the Boundry Creation dialog box below. Click on the Pick Points button in the upper right corner of the dialog box. Then pick an internal point to automatically trace a polyline around a closed shape in the color of the current layer. It must be emphasized that the shape must be completely closed for the BOUNDARY command to function properly.

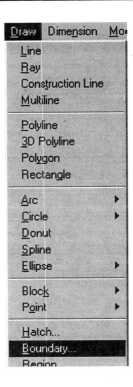

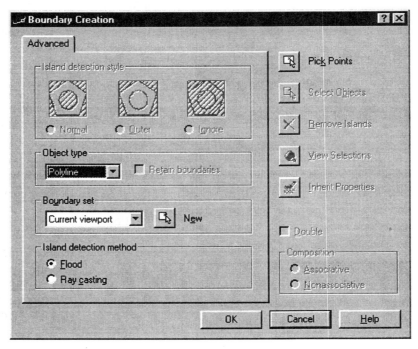

In the illustration below of the Extrusion, issuing the BOUNDARY command and clicking on the Pick Points< button of the dialog box prompts the user to pick an internal point. Selecting a point inside of the Extrusion at "A" traces the polyline on the current layer. Turning Off the layer containing the individual objects leaves the polyline on which you can perform various calculations.

Command: **BO** *(For BOUNDARY)*
Select internal point: *(Select a point inside of the extrusion at "A")*
Selecting everything...
Selecting everything visible...
Analyzing the selected data...
Analyzing internal islands...
Select internal point: *(Press ENTER to create the boundary polyline on the Boundary layer)*
BOUNDARY created 1 polyline

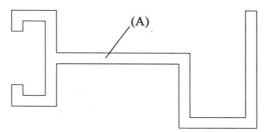

(A)

Step #3

With the entire extrusion converted into a polyline, use the FILLET command, set a radius of 0.050, and use the Polyline option of the FILLET command to fillet all corners of the extrusion at once.

Command: **F** *(For FILLET)*
Current settings: Mode = TRIM, Radius = 0.5000
Select first object or [Polyline/Radius/Trim]: **R** *(For Radius)*
Specify fillet radius <0.5000>: **0.050**

Command: **F** *(For FILLET)*
Current settings: Mode = TRIM, Radius = 0.0500

Select first object or [Polyline/Radius/Trim]: **P** *(For Polyline)*
Select 2D polyline: *(Select the polyline at the right)*
22 lines were filleted

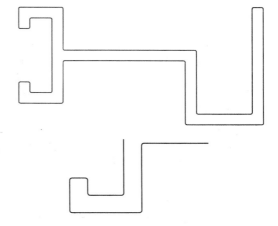

Checking the Accuracy of Extrude.Dwg

Once the extrusion has been constructed, answer the question below to determine the accuracy of the drawing.

Question #1

What is the total surface area of the extrusion?

      (A) 7.020
      (B) 7.070
      (C) 7.120
      (D) 7.170

Use the AREA command to calculate the surface area of the extrusion. This is easily accomplished since the extrusion has already been converted into a polyline.

Command: **AA** *(For AREA)*
Specify first corner point or [Object/
    Add/Subtract]: **O** *(For Object)*
Select objects: *(Select any part of
    the extrusion)*
Area = 7.170, Perimeter = 38.528

Total surface area of the extrusion is "D", 7.170

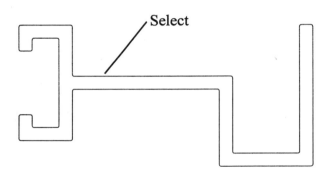

Select

Tutorial Exercise
C-Lever.Dwg

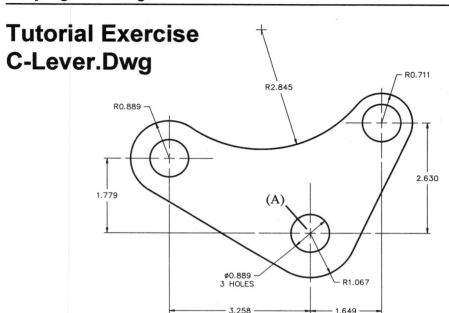

Purpose

This tutorial is designed to show you various methods to construct the C-Lever object above. Numerous questions will be asked about the object requiring the use of the AREA, DIST, ID, and LIST commands.

System Settings

Use the Drawing Units dialog box and change the number of decimal places past the zero from four units to three units. Keep the default drawing limits set at 0.000,0.000 and 12.000,9.000. Check to see that the following Object Snap modes are already set: Endpoint, Extension, Intersection, Center

Layers

Create the following layers:

Name	Color	Linetype
Boundary	Magenta	Continuous
Object	Yellow	Continuous

Suggested Commands

Begin drawing the C-Lever with point "A" illustrated above at absolute coordinate 7.000,3.375. Begin laying out all circles. Then draw tangent lines and arcs. Use the TRIM command to clean up unnecessary objects. To prepare to answer the AREA command question, convert the profile of the C-Lever into a polyline using the BOUNDARY command. Other questions pertaining to distances, angles, and point identifications follow.

Step #1

Make the Object layer current. Then, construct one circle of 0.889 diameter with the center of the circle at absolute coordinate 7.000,3.375 (see Figure 13–18). Construct the remaining circles of the same diameter by using the COPY command with the Multiple option. Use of the @ symbol for the base point in the COPY command identifies the last known point, which in this case is the center of the first circle drawn at coordinate 7.000,3.375.

Command: **C** *(For CIRCLE)*
Specify center point for circle or [3P/2P/Ttr (tan tan radius)]: **7.000,3.375**
Specify radius of circle or [Diameter]: **D** *(For Diameter)*

Specify diameter of circle: **0.889**

Command: **CP** *(For COPY)*
Select objects: **L** *(For Last)*
Select objects: *(Press ENTER to continue)*
Specify base point or displacement, or [Multiple]: **M** *(For Multiple)*
Specify base point: **@**
Specify second point of displacement or <use first point as displacement>: **@1.649,2.630**
Specify second point of displacement or <use first point as displacement>: **@-3.258,1.779**
Specify second point of displacement or <use first point as displacement>: *(Press ENTER to exit this command)*

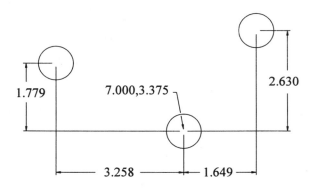

Step #2

Construct three more circles. Even though these objects actually represent arcs, circles will be drawn now and trimmed later to form the arcs.

Command: **C** *(For CIRCLE)*
Specify center point for circle or [3P/2P/Ttr (tan tan radius)]: *(Select the edge of the circle at "A" to snap to its center)*
Specify radius of circle or [Diameter] <0.445>: **1.067**

Command: **C** *(For CIRCLE)*
Specify center point for circle or [3P/2P/Ttr (tan tan radius)]: *(Select the edge of the circle at "B" to snap to its center)*
Specify radius of circle or [Diameter] <1.067>: **0.889**

Command: **C** *(For CIRCLE)*
Specify center point for circle or [3P/2P/Ttr (tan tan radius)]: *(Select the edge of the circle at "C" to snap to its center)*
Specify radius of circle or [Diameter] <0.889>: **0.711**

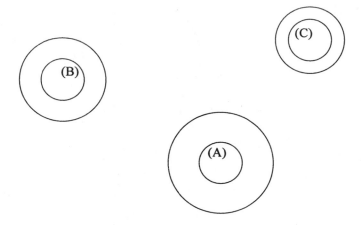

Step #3

Construct lines tangent to the three outer circles illustrated at the right.

Command: **L** *(For LINE)*
Specify first point: **Tan**
to *(Select the outer circle near "A")*
Specify next point or [Undo]: **Tan**
to *(Select the outer circle near "B")*
Specify next point or [Undo]: *(Press
   ENTER to exit this command)*

Command: **L** *(For LINE)*
Specify first point: **Tan**
to *(Select the outer circle near "C")*
Specify next point or [Undo]: **Tan**
to *(Select the outer circle near "D")*
Specify next point or [Undo]: *(Press
   ENTER to exit this command)*
To point: *(Strike ENTER to exit this
command.)*

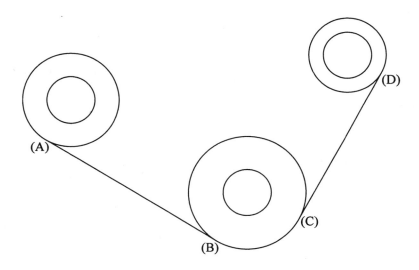

Step #4

Construct a circle tangent to the two circles illustrated at the right using the CIRCLE command with the Tangent-Tangent-Radius option (TTR).

Command: **C** *(For CIRCLE)*
Specify center point for circle or [3P/2P/Ttr (tan tan radius)]: **TTR**
Specify point on object for first tangent of circle: *(Select the outer circle near "A")*
Specify point on object for second tangent of circle: *(Select the outer circle near "B")*
Specify radius of circle <0.711>: **2.845**

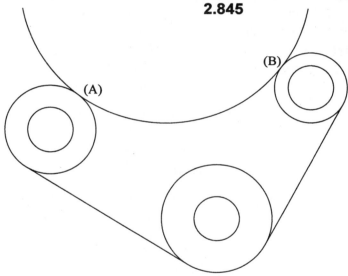

Step #5

Use the TRIM command to clean up and form the finished drawing. Select all of the objects represented by dashed lines as cutting edges. Follow the prompts below for selecting the objects to trim.

Command: **TR** *(For TRIM)*
Current settings: Projection=UCS Edge=None
Select cutting edges ...
Select objects: (Select all dashed objects illustrated below)
Select objects: *(Press ENTER to continue)*

Select object to trim or [Project/ Edge/Undo]: *(Select the circle at "A")*
Select object to trim or [Project/ Edge/Undo]: *(Select the circle at "B")*
Select object to trim or [Project/ Edge/Undo]: *(Select the circle at "C")*
Select object to trim or [Project/ Edge/Undo]: *(Select the circle at "D")*
Select object to trim or [Project/ Edge/Undo]: *(Press ENTER to exit this command)*

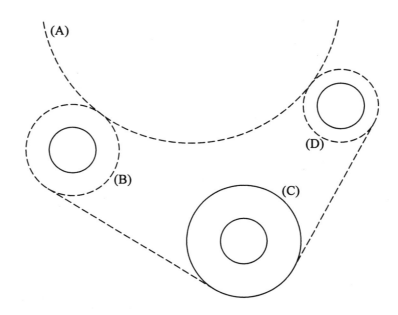

Checking the Accuracy of C-Lever.Dwg

Once the C-Lever has been constructed, answer the questions below to determine the accuracy of this drawing. Use the illustration above to assist in answering the questions.

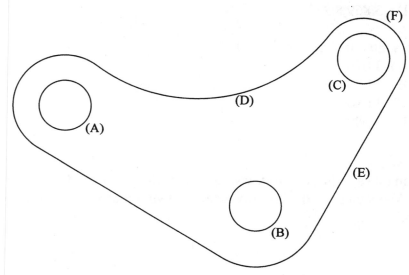

1. What is the total area of the C-Lever with all three holes removed?
   - (A) 13.744
   - (B) 13.749
   - (C) 13.754
   - (D) 13.759

2. What is the total distance from the center of circle "A" to the center of circle "B"?
   - (A) 3.702
   - (B) 3.707
   - (C) 3.712
   - (D) 3.717

3. What is the angle formed in the X-Y plane from the center of circle "C" to the center of circle "B"?
   - (A) 223 degrees
   - (B) 228 degrees
   - (C) 233 degrees
   - (D) 238 degrees

4. What are the delta X-Y distances from the center of circle "C" to the center of circle "A"?
   - (A) -4.907,-0.851
   - (B) -4.907,-0.856
   - (C) -4.907,-0.861
   - (D) -4.907,-0.866

5. What is the absolute coordinate value of the center of arc "D"?
   - (A) 5.869,8.218
   - (B) 5.869,8.223
   - (C) 5.869,8.228
   - (D) 5.869,8.233

6. What is the total length of line "E"?
   - (A) 3.074
   - (B) 3.079
   - (C) 3.084
   - (D) 3.089

7. What is the total length of arc "F"?
   - (A) 2.071
   - (B) 2.076
   - (C) 2.081
   - (D) 2.086

A solution for each question follows, complete with the method used to arrive at the answer. Apply these methods to any type of drawing that requires the use of Inquiry commands.

Question #1

What is the total area of the C-Lever
with all three holes removed?
>   (A) 13.744
>   (B) 13.749
>   (C) 13.754
>   (D) 13.759

First make the Boundary layer current.
Then use the BOUNDARY command
and pick a point inside of the object at
"A". This will trace a polyline around
all closed objects on the Boundary layer.

Command: **BO** *(For BOUNDARY)*
Select internal point: *(Pick a point
  inside of the object at "Y")*
Selecting everything...
Selecting everything visible...
Analyzing the selected data...
Analyzing internal islands...
Select internal point: *(Press EN-
  TER to create the boundaries)*
BOUNDARY created 4 polylines

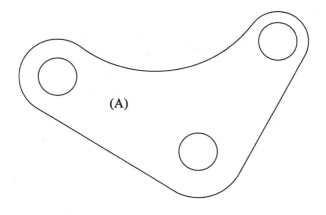

(A)

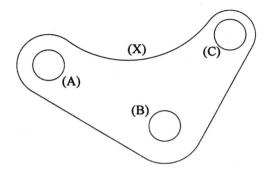

Next, turn off the Object layer. All objects on the Boundary layer should be visible. Then use the AREA command to add and subtract objects to arrive at the final area of the object.

Command: **AA** *(For AREA)*
Specify first corner point or [Object/ Add/Subtract]: **A** *(For Add)*
Specify first corner point or [Object/ Subtract]: **O** *(For Object)*
(ADD mode) Select objects: *(Select the edge of the shape near "X")*
Area = 15.611, Perimeter = 17.771
Total area = 15.611
(ADD mode) Select objects: *(Press ENTER to continue)*
Specify first corner point or [Object/ Subtract]: **S** *(For Subtract)*
Specify first corner point or [Object/ Add]: **O** *(For Object)*

(SUBTRACT mode) Select objects: *(Select circle "A")*
Area = 0.621, Perimeter = 2.793
Total area = 14.991
(SUBTRACT mode) Select objects: *(Select circle "B")*
Area = 0.621, Perimeter = 2.793
Total area = 14.370
(SUBTRACT mode) Select objects: *(Select circle "C")*
Area = 0.621, Perimeter = 2.793
Total area = 13.749
(SUBTRACT mode) Select objects: *(Press ENTER to continue)*
Specify first corner point or [Object/ Add]: *(Press ENTER to exit this command)*

The total area of the C-Lever with all three holes removed is (B), 13.749

Question #2

What is the total distance from the center of circle "A" to the center of circle "B"?

    (A) 3.702
    (B) 3.707
    (C) 3.712
    (D) 3.717

Use the DIST (Distance) command to calculate the distance from the center of circle "A" to the center of circle "B" in the figure below. Be sure to use the OSNAP-Center mode for locating the centers of all circles.

Command: **DI** *(For DIST)*
Specify first point: *(Select the edge of the circle at "A")*
Specify second point: *(Select the edge of the circle at "B")*
Distance = 3.712
Angle in XY Plane = 331
Angle from XY Plane = 0
Delta X = 3.258, Delta Y = -1.779, Delta Z = 0.000

The total distance from the center of circle "A" to the center of circle "B" is (C), **3.712.**

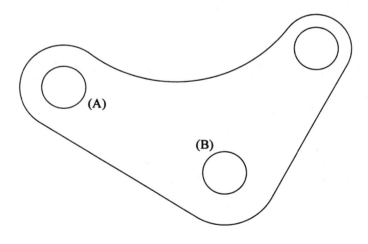

Question #3

What is the angle formed in the X-Y plane from the center of circle "C" to the center of circle "B"?

      (A) 223 degrees
      (B) 228 degrees
      (C) 233 degrees
      (D) 238 degrees

Use the DIST (Distance) command to calculate the angle from the center of circle "C" to the center of circle "B" in the figure below. Be sure to use the OSNAP-Center mode for locating the centers of all circles.

Command: **DI** *(For DIST)*
Specify first point: *(Select the edge of the circle at "C")*
Specify second point: *(Select the edge of the circle at "B")*
Distance = 3.104
Angle in XY Plane = 238
Angle from XY Plane = 0
Delta X = -1.649, Delta Y = -2.630, Delta Z = 0.000

The angle formed in the X-Y plane from the center of circle "C" to the center of circle "B" is (D), <u>238 degrees.</u>

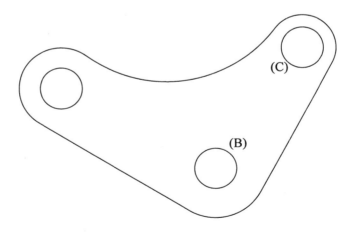

Question #4

What is the delta X-Y distances from the center of circle "C" to the center of circle "A"?

     (A) -4.907,-0.851
     (B) -4.907,-0.856
     (C) -4.907,-0.861
     (D) -4.907,-0.866

Use the DIST (Distance) command to calculate the delta X,Y distance from the center of circle "C" to the center of circle "A" in the figure below. Be sure to use the OSNAP-Center mode. Notice that additional information is given when you use the DIST command. For the purpose of this question, we will only be looking for the delta X,Y distance. The DIST command will display the relative X,Y,Z distances. Since this is a 2D problem, only the X and Y values will be used.

Command: **DI** *(For DIST)*
Specify first point: *(Select the edge of the circle at "C")*
Specify second point: *(Select the edge of the circle at "A")*
Distance = 4.980
Angle in XY Plane = 190
Angle from XY Plane = 0
Delta X = -4.907, Delta Y = -0.851,
  Delta Z = 0.000

The delta X-Y distance from the center of circle "C" to the center of circle "A" is (A), -4.907, -0.851.

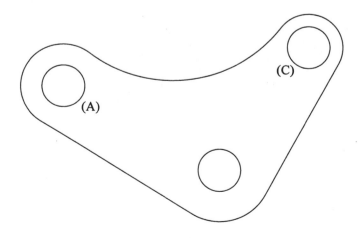

Question #5

What is the absolute coordinate value of the center of arc "D"?

(A) 5.869,8.218
(B) 5.869,8.223
(C) 5.869,8.228
(D) 5.869,8.233

The ID command is used to get the current absolute coordinate information on a desired point (see the figure below). This command will display the X,Y,Z coordinate values. Since this is a 2D problem, only the X and Y values will be used.

Command: **ID**
Specify point: *(Select the edge of the arc at "D"; OSNAP Center mode should be active)*
X = 5.869   Y = 8.223   Z = 0.000

The absolute coordinate value of the center of arc "D" is (B), 5.869,8.223.

?
+

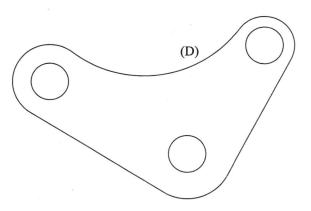

(D)

Question #6

What is the total length of line "E"?
- **(A) 3.074**
- **(B) 3.079**
- **(C) 3.084**
- **(D) 3.089**

Use the DIST (Distance) command to find the total length of line "E" in the figure below. Be sure to use the OSNAP-Endpoint mode. Notice that additional information is given when using the DIST command. For the purpose of this question, we will only be looking for the distance.

Command: **DI** *(For DIST)*
Specify first point: (Select the endpoint of the line at "X")
Specify second point: (Select the endpoint of the line at "Y")
Distance = 3.084
Angle in XY Plane = 64
Angle from XY Plane = 0
Delta X = 1.328, Delta Y = 2.783, Delta Z = 0.000

The total length of line "E" is (C), 3.084.

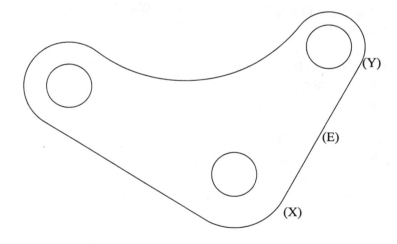

Question #7

What is the total length of arc "F"?
- **(A) 2.071**
- **(B) 2.076**
- **(C) 2.081**
- **(D) 2.086**

The LIST is used to calculate the lengths of arcs. However, a little preparation is needed before performing this operation. If arc "F" is selected as in the figure below, notice that the entire outline is selected because it is a polyline. Use the EXPLODE command to break the outline back into individual objects. Use the LIST command to get a listing of the arc length.

Command: **X** *(For EXPLODE)*
Select objects: *(Select the edge of the dashed polyline in the figure below)*
Select objects: *(Press ENTER to perform the explode operation)*

Command: **LI** *(For LIST)*
Select objects: *(Select the edge of the arc at "F" in the figure below)*
Select objects: *(Press ENTER to continue)*

ARC   Layer: "Boundary"
Space: Model space
Handle = 48
center point, X=8.649   Y=6.005
Z=0.000
radius   0.711
start angle   334
end angle   141
length   2.071

The total length of arc "F" is (A), 2.071.

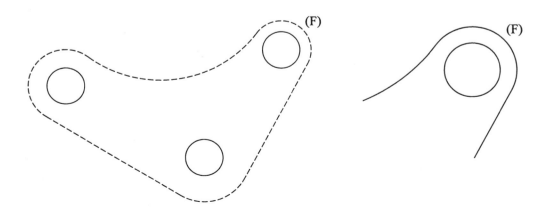

Chapter 4

AutoCAD 2000 Level I Practice Test

This AutoCAD 2000 Level I Practice Test has been developed to provide you with more experience in preparing for the AutoCAD Level I Assessment Exam. This practice test consists of 50 drawing and general knowledge questions. There is no time limit to complete this assessment test. However, this test is designed to be completed in 2 hours or less, which would demonstrate use of the software in a productive manner.

Question types include single answer multiple choice and hot spot areas. Numerous questions have been designed around actual images an individual would be confronted with in the production drawing environment.

Two types of drawings are present in this practice test. Most of the drawings are already created up to a certain point. For these cases, open the drawing file and follow the steps that direct you to perform certain operations before attempting to answer any of the questions that relate to the drawing. All drawings to open are provided on the CD supplied with this manual. Create a folder called \ASSESS-MENT and load all drawing files there. Another type of drawing requires you to construct a new object from the image provided and answer the questions that follow the drawing to test your accuracy.

Answers for all Level I Practice Test questions are located at the end of this chapter.

Notes

Provide the best answer for each of the following questions.

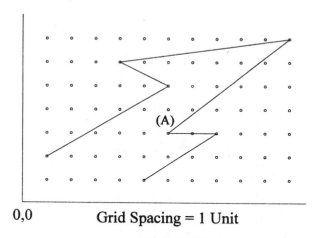

0,0   Grid Spacing = 1 Unit

1. In the illustration above, what is the the absolute coordinate value of Point "A"?
    (A) 6,3
    (B) 3,6
    (C) @6<30
    (D) @3<90
Level I Objective 3.01

2. What is the correct Relative coordinate listing to construct a line segment 6 units
    in the X-direction and 11 units in the Y-direction?
    (A) 11,6
    (B) 6,11
    (C) @6,11
    (D) @11,6
Level I Objective 3.02

3. What mode allows you to control the construction of lines at angles other than
    90 degrees?
    (A) Osnap
    (B) Ortho
    (C) Polar Tracking
    (D) Object Tracking
Level I Objective 3.05

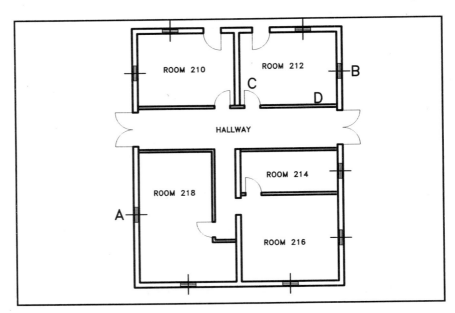

4. Open the existing drawing called BLOCK WALLS. Stretch the bottom wall of this floor plan 2'-0" straight down. Turn off all windows and center lines on this drawing. What is the total area of all thick block walls?

    (A) 193 sq. ft.
    (B) 197 sq. ft.
    (C) 201 sq. ft.
    (D) 205 sq. ft.

Level I Objective 1.02

5. You use a combination of Window and Fence to build a selection set of objects and perform a move operation. You now want to create a copy of this same selection set. What selection set mode is used to retrieve this selection set without recreating it?

    (A) All
    (B) Last
    (C) Previous
    (D) Window Polygon

Level I Objective 5.02

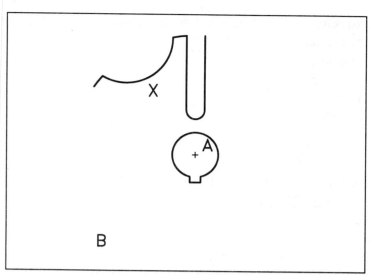

6. Open the existing drawing called GENEVA. Use the ARRAY command to copy the shape marked "X" a total of six times in a full circular pattern. Use the center of arc "A" as the center of the array. What is the total area of the shape with the hole and keyway?

    (A) 27.37
    (B) 27.42
    (C) 27.47
    (D) 27.52

Level I Objective 5.06

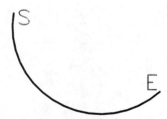

7. You construct an arc with the "S,E,C" option as in the illustration above. What does the letter "C" stands for?

    (A) Center
    (B) Circumference
    (C) Clockwise
    (D) Counterclockwise

Level I Objective 3.08

8. You need to select multiple lines to trim to a cutting edge. What selection set
    mode is used in this operation?
    (A) Crossing
    (B) Crossing Polygon
    (C) Fence
    (D) Window

Level I Objective 5.01

9. From the following lists, what objects can be used as valid boundary edges
    when using the EXTEND command?
    (A) circles, lines, and arcs.
    (B) blocks, circles, and lines.
    (C) unexploded hatch patterns, arcs, and lines.
    (D) polylines, unexploded hatch patterns, and circles.

Level I Objective 5.11

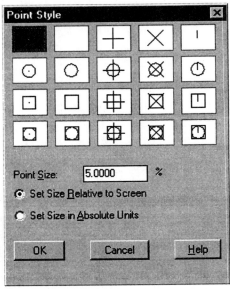

10. Click in the area of the Point Style dialog box designed to set the point display
    to a fixed size.

Level I Objective 3.09

11. Grip mode is currently turned on. You pick an object and grips appear on the object. Which are valid grip modes?

    (A) Trim, Mirror, and Erase

    (B) Rotate, Scale, and Move

    (C) Break, Stretch, and Trim

    (D) Move, Break, and Extend

Level I Objective 5.15

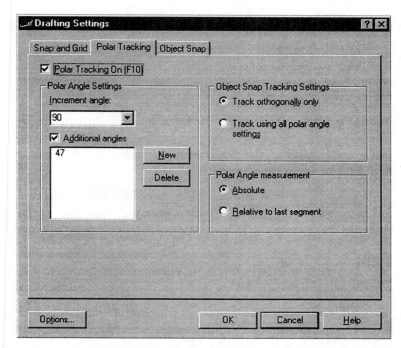

12. Click on the area in the Polar Tracking tab designed to snap to only the specified angle. Multiples of that angle are not supported in this area.

Level I Objective 3.06

13. You have acquired a point to be used for Object Snap Tracking. This, however, was the wrong point to be used. How do you clear the acquired point?

    (A) Press the ESC key.

    (B) Pick the acquired point.

    (C) Move your cursor over the acquired point and pause.

    (D) While holding down the SHIFT key, pick the acquired point.

Level I Objective 3.12

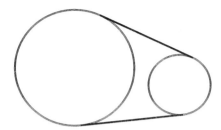

14. Two circles are constructed in the illustration above. Two line segments need to be constructed to the edges of both circles. What Object Snap mode is used to accomplish this construction task?

    (A) Center
    (B) Edge
    (C) Quadrant
    (D) Tangent

Level I Objective 3.11

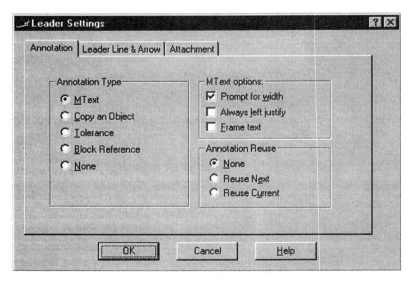

15. Click on the area of the Leader Settings dialog box designed to take you to the area for setting a spline object as the leader.

Level I Objective 7.02

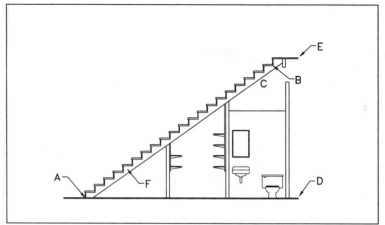

16. Open the existing drawing called STAIRS. Divide the bottom of the stringer identified by line "C" into seven equal parts. What is the distance from the point created at "C" perpendicular to the top of the finished floor identified by "D"?

    (A) 10'-7"
    (B) 10'-10"
    (C) 11'-1"
    (D) 11'-4"

Level I Objective 3.10

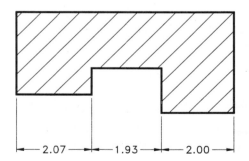

17. You activate the QDIM command. Dimensions need to be efficiently placed side-by-side as in the illustration above. What option of the QDIM command would you use to accomplish this task?

    (A) Baseline
    (B) Continuous
    (C) Ordinate
    (D) Staggered

Level I Objective 7.01

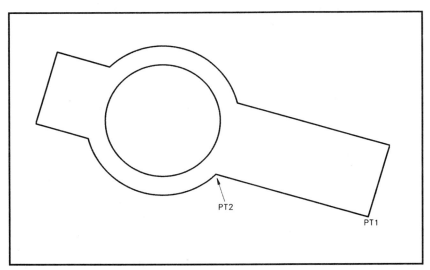

18. Open the existing drawing called SCALE2. Increase the size of all yellow objects so the distance between PT1 and PT2 is 5 units. What is the total area of the shape minus the hole?

    (A) 23.3561
    (B) 23.3565
    (C) 23.3569
    (D) 23.3573

Level I Objective 5.09

19. You need to place text in a drawing. The text must fit in an area so that when the text is entered through a dialog box, it automatically wraps to the next line. What command is used to accomplish this task?

    (A) DTEXT
    (B) MTEXT
    (C) TEXT
    (D) TXT

Level I Objective 6.02

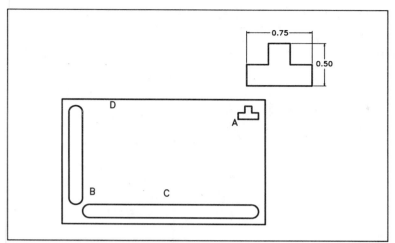

20. Open the existing drawing called PLATE. Move this entire drawing 4.38 units in the X direction and 6.88 units in the Y direction. What is the new XY coordinate location of the intersection at "A"?

    (A) 11.53,11.31
    (B) 11.59,11.37
    (C) 11.59,11.43
    (D) 11.65,11.49

Level I Objective 5.04

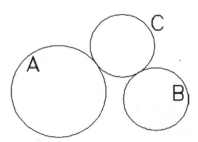

21. You construct circle "C" tangent to circles "A" and "B" in the illustration above. What option of the CIRCLE command is designed to accomplish this operation?

    (A) 2P
    (B) 3P
    (C) TTR
    (D) TTT

Level I Objective 3.07

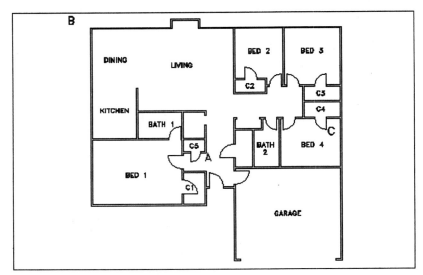

22. Open the existing drawing called FIRST FLOOR. Rotate the floor plan counterclockwise 90 degrees. Use the intersection of the corner at "A" as the base point of rotation. What is the X-Y coordinate value of the intersection at the outer corner identified by "B"?

    (A) -2'-7",39'-8"
    (B) -2'-7",40'-0"
    (C) -2'-11",40'-0"
    (D) -2'-11",40'-4"

Level I Objective 5.08

23. You are in the process of selecting one object to edit in a very complicated drawing. While in the "Select objects:" prompt, you press the CTRL key. What function does this action allow?

    (A) Window
    (B) Object Cycling
    (C) Window Polygon
    (D) Crossing Polygon

Level I Objective 5.03

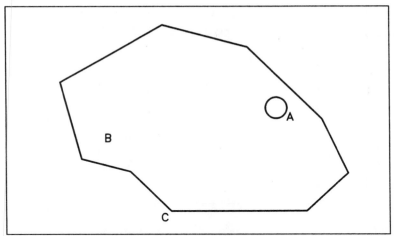

24. Open the existing drawing called PATTERN TEMPLATE. Stretch vertex "C" a distance of 10 units down and 30 units to the right. What is the new angular measurement of vertex "C"?

    (A) 135 degrees
    (B) 138 degrees
    (C) 141 degrees
    (D) 144 degrees

Level I Objective 5.10

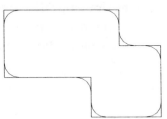

25. You add fillets to an object and the results are illustrated above. What mode of the FILLET command is used to place the fillet and leave the line segments untrimmed?

    (A) Trim
    (B) No Edit
    (C) No Trim
    (D) Leave Edge

Level I Objective 5.13

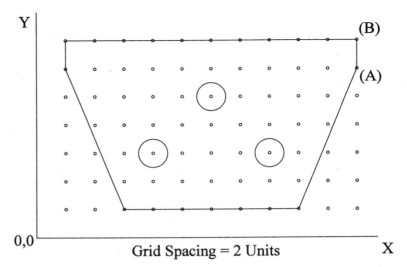

Grid Spacing = 2 Units

26. In the illustration above, what could you enter at the keyboard to draw a line from Point "B" to Point "A"?

(A) @2.00<0
(B) @2.00<90
(C) @2.00<180
(D) @2.00<270

Level I Objective 3.03

27. As a means of compressing the drawing file to increase drawing productivity, you eliminate unused blocks, dimension styles, layers, layouts, linetypes, and text styles from the current drawing. What command is used to perform this operation?

(A) COMPRESS
(B) DELETE
(C) ERASE
(D) PURGE

Level I Objective 8.04

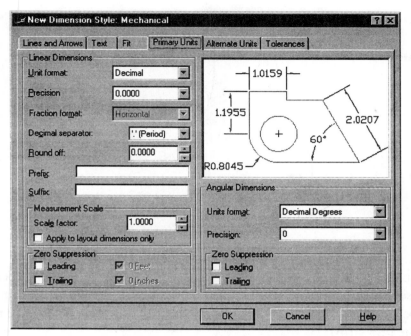

28. Using the illustration above, click on the area of the Primary Units tab of the New Dimension Style dialog box designed to change the number of decimal places for dimension text.
Level I Objective 7.03

29. You need to create a symbol of a door that is to be inserted into numerous drawing files. What command is used to create a global symbol?
   (A) BLOCK
   (B) DITTO
   (C) SYMBOL
   (D) WBLOCK
Level I Objective 8.01

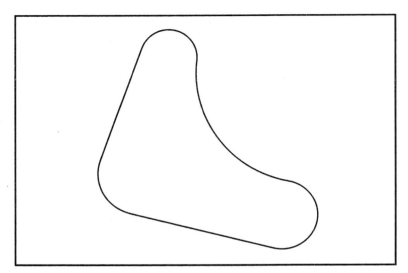

30. Open the existing drawing called BRACKET SLIDE. Turn on layer Object. Offset the perimeter of Bracket Slide. Use 0.573 units as the distance of the offset. Pick inside of Bracket Slide as the side to offset. What is the total area of the thin strip?

    (A) 13.283
    (B) 13.289
    (C) 13.295
    (D) 13.301

Level I Objective 5.05

31. You have constructed a line segment. You now need to split the line segment into two equal segments. You must accomplish this without seeing where the line segments were split. What command can be used to accomplish this task?

    (A) BREAK
    (B) ERASE
    (C) SLICE
    (D) TRIM

Level I Objective 5.14

Start a new drawing called Pattern4. Keep the default setting of decimal units but change the number of decimal places past the zero from 4 to 0. Begin constructing Pattern4 with vertex "A" at absolute coordinate (50,30). Dimensions do not have to be added to this drawing. When finished, answer Question #32.

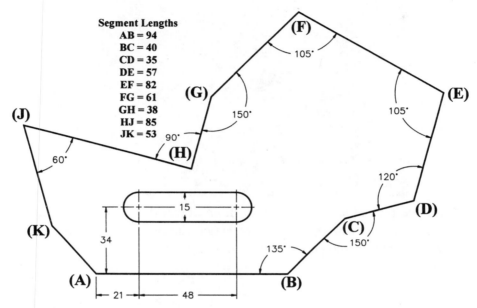

Segment Lengths
AB = 94
BC = 40
CD = 35
DE = 57
EF = 82
FG = 61
GH = 38
HJ = 85
JK = 53

32. After completing the drawing of Pattern4 illustrated above, what is the total area of Pattern4 with the slot removed?

    (A) 14501
    (B) 14539
    (C) 14561
    (D) 14596

Level I Objective 4.01

33. What value would you enter to construct a line 10' long in the 45 degree direction using the Direct Distance mode?

    (A) 10'
    (B) @10',10'
    (C) @10'<45
    (D) @10'<-45

Level I Objective 3.04

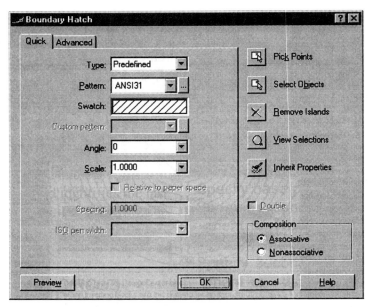

34. You have picked an internal point in an object designed to be crosshatched. Click in the area of the Boundary Hatch dialog box designed to check the accuracy of the hatch pattern before making the pattern a permanent part of the drawing database.

Level I Objective 6.05

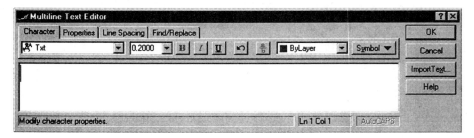

35. You have created a document in Word and saved it in a RTF format. Click on the proper area of the Multiline Text Editor dialog box designed to merge this RTF file into your drawing.

Level I Objective 6.03

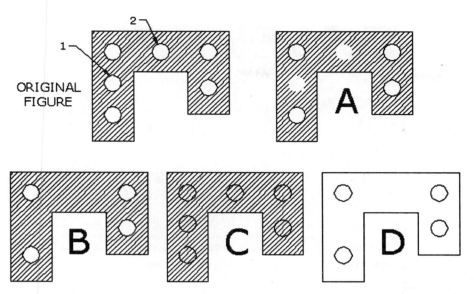

36. In the original figure of the object illustrated above, an associative crosshatch pattern is placed in the object. Circles "1" and "2" are deleted from the object. Which view best describes the results of this editing operation?
    (A) A
    (B) B
    (C) C
    (D) D

Level I Objective 6.06

37. What option of the ZOOM command allows you to get back to the last zoomed display in the quickest way possible?
    (A) Center
    (B) Last
    (C) Previous
    (D) Window

Level I Objective 2.01

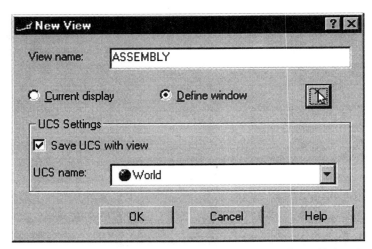

38. What does the "Define window" option of the New View dialog box illustrated above allow you to do?

    (A) Change to a new viewing point.

    (B) Save a view by means of a window.

    (C) Restore a view by means of a window.

    (D) Call up a view listing within a window.

Level I Objective 2.02

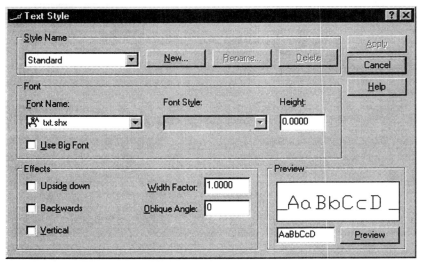

39. Click on the area of the Text Style dialog box to change to a different font.

Level I Objective 6.01

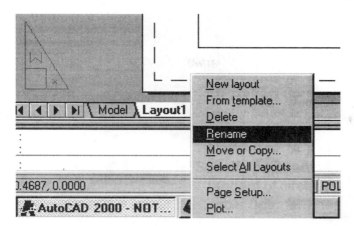

40. What action displays the menu illustrated above which is designed to rename "Layout1"?

    (A) Left clicking on the Layout1 tab.

    (B) Right clicking on the Layout1 tab.

    (C) Double clicking on the Layout1 tab.

    (D) Holding down the SHIFT key while left clicking on the Layout1 tab.

Level I Objective 9.03

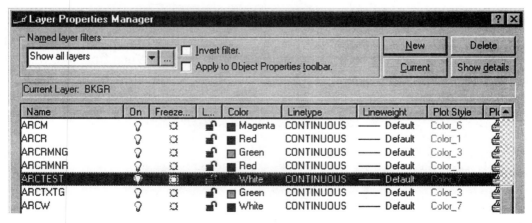

41. You have mistakenly created the layer ARCTEST in the Layer Properties Manager dialog box above. Click on the area of this dialog box designed to remove this layer from the list.

Level I Objective 1.01

42. Click on the button in the illustration above that will launch the Properties window.

Level I Objective 5.17

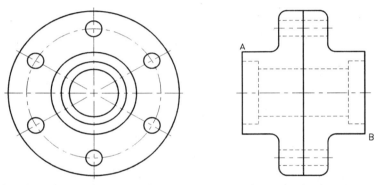

43. Open the drawing called COUPLING. Use Line1 and Line 2 as mirror lines to create the right side view in the illustration above. When finished, what is the distance from the intersection at "A" to the intersection at "B"?

    (A) 4.5763
    (B) 4.5767
    (C) 4.5771
    (D) 4.5775

Level I Objective 5.07

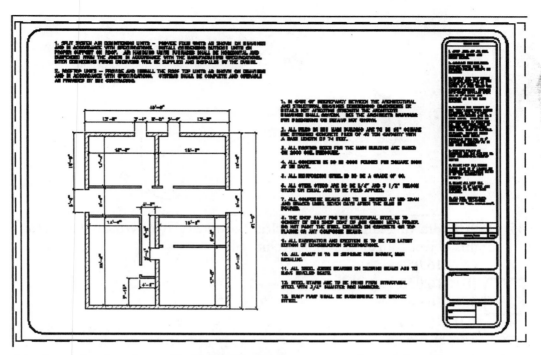

44. Open the drawing called BUILDING PLAN 2. After performing a search on all text, what is the name of misspelled word?

    (A) BARING
    (B) MAS
    (C) ROLER
    (D) TRUS

Level I Objective 6.04

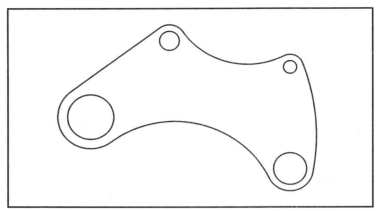

45. Open the drawing called CAM which is illustrated above. What is the total area of this object with all four holes removed?

(A) 116.054
(B) 116.058
(C) 116.062
(D) 116.066

Level I Objective 5.16

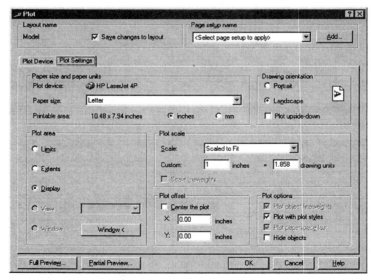

46. Click on the area of the Plot dialog box designed to plot the area of a drawing based on the objects that make up the drawing.

Level I Objective 9.01

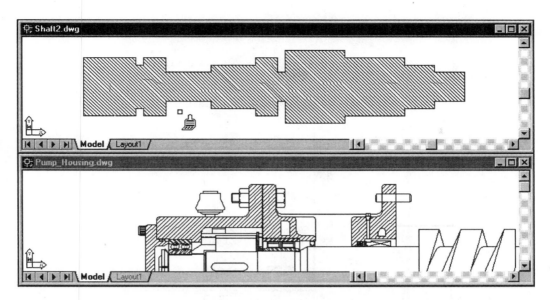

47. You open up two drawings in a Multiple Drawing Environment. In the illustration above, what operation is signified by the appearance of the paintbrush icon in the upper viewport?

    (A) Copy

    (B) Move

    (C) Match Property

    (D) Drag and Drop as a Block

Level I Objective 8.02

48. What command is used to delete portions of objects based on a cutting edge?

    (A) BREAK

    (B) ERASE

    (C) REMOVE

    (D) TRIM

Level I Objective 5.12

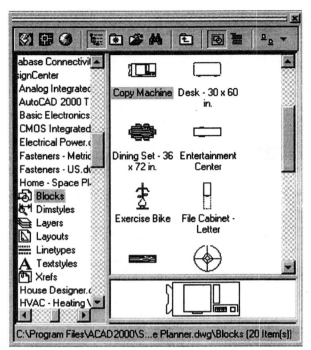

49. Click on the area of the DesignCenter dialog box designed to produce an enlarged image of a selected block.

Level I Objective 8.03

50. You have produced a full size floor plan drawing in model space. You have arranged this drawing in a layout in paper space by zooming to the proper scale in the floating viewport. At what scale should this layout be plotted?

(A) 1=1
(B) 1/4=1'-0"
(C) 1=10
(D) Fit

Level I Objective 9.02

Answers to the AutoCAD 2000 Level I Practice Test

1. A

2. C

3. C

4. C

5. C

6. D

7. A

8. C

9. A

10. Click in the radio button adjacent to "Set Size in Absolute Units" in the Point Style dialog box.

11. B

12. Place a check in the box adjacent to "Additional angles".

13. C

14. D

15. Click on the Leader Line & Arrow tab to change to a spline leader.

16. A

17. B

18. D

19. B

20. A

21. C

22. C

23. B

24. B

25. C

26. D

27. D

28. In the Linear Dimensions area, click in the box adjacent to "Precision".

29. D

30. B

31. A

32. B

33. A

34. Click on the Preview button.

35. Click on the Import Text button.

36. B

37. C

38. B

39. Click in the "Font Name:" box to change the name of the font.

40. B

41. Click on the Delete button to remove the highlighted layer

42. Click on this button:

47. C

48. D

43. D

49. Click on this button:

44. C

45. B

50. A

46. In Plot area, click on the radio button adjacent to "Extents".

Chapter 5

AutoCAD 2000 Level I Exit Exam

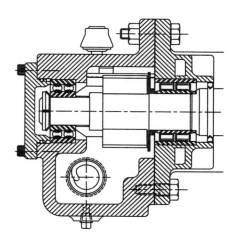

This AutoCAD 2000 Level I Exit Exam has been developed for more practice in preparing for the AutoCAD Level I Assessment Exam. The Exit Exam consists of 50 drawing and general knowledge questions. There is no time limit to complete this assessment exam. However, this exam is designed to be completed in 2 hours or less, which would demonstrate use of the software in a productive manner.

Question types include single answer multiple choice and hot spot areas. Numerous questions have been designed around actual images an individual would be confronted with in the production drawing environment.

Two types of drawings are present in this exit exam. Most of the drawings are already created up to a certain point. For these cases, open the drawing file and follow the steps that direct you to perform certain operations before attempting to answer any of the questions that relate to the drawing. All drawings to open are provided on the disk supplied with this manual. Create a new folder called \ASSESSMENT and load all drawing files there. Another type of drawing requires you to construct a new object from the image provided and answer the questions that follow the drawing to test your accuracy.

Work through this Exit Exam at a good pace, paying strict attention to the amount of time spent on each question. Answers for all Level I Exit Exam questions are located at the end of this chapter.

Notes

Provide the best answer for each of the following questions.

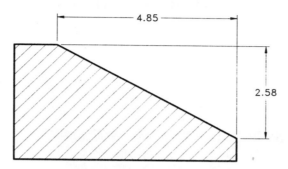

1. In the illustration above, what command is used to place either a horizontal or vertical dimension?

  (A) DIMHORIZONTAL
  (B) DIMVERTICAL
  (C) DIMLINEAR
  (D) LINEAR

Level I Objective 7.01

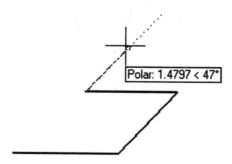

2. What mode has been activated in the image provided?

  (A) Ortho
  (B) Polar Tracking
  (C) Polar Coordinates
  (D) Relative Coordinates

Level I Objective 3.05

3. You are in a busy drawing. You need to use Object Cycling as a means of cycling through various objects until the correct object is selected. What key is pressed at the "Select objects:" prompt designed to activate Object Cycling?

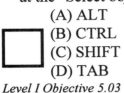

    (A) ALT
    (B) CTRL
    (C) SHIFT
    (D) TAB

Level I Objective 5.03

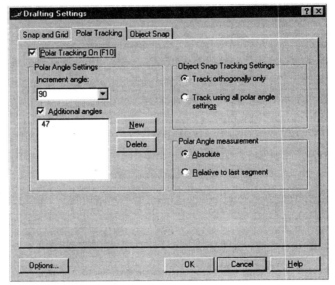

4. Click on the area in the Polar Tracking tab that will display Polar Tracking angles based on the angle of the last two points selected.

Level I Objective 3.06

5. Which function key is used to turn on Object Snap Tracking?

    (A) F8
    (B) F9
    (C) F10
    (D) F11

Level I Objective 3.12

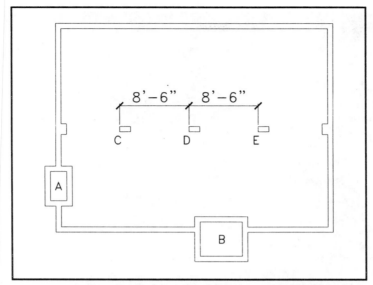

6. Open the existing drawing called FOUNDATION. The concrete block walls around "A" and "B" are 8". Copy concrete pier "C" to locations "D" and "E" using the distances illustrated above. What is the total area of all concrete block walls and the three piers?

    (A) 96 sq. ft.
    (B) 99 sq. ft.
    (C) 102 sq. ft.
    (D) 105 sq. ft.

Level I Objective 5.04

7. You create a point on your screen using the POINT command. However, the point is difficult to see since it has the appearance of a dot. You change to a different point style in the Point Style dialog box. When clicking the OK button to exit the dialog box, what occurs?

    (A) The new point style immediately displays on the screen.
    (B) Nothing until you perform a ZOOM All. The new point style appears.
    (C) Nothing. You must first redraw the screen before the new point style can be displayed.
    (D) Nothing. You must first regenerate the screen before the new point style can be displayed.

Level I Objective 3.09

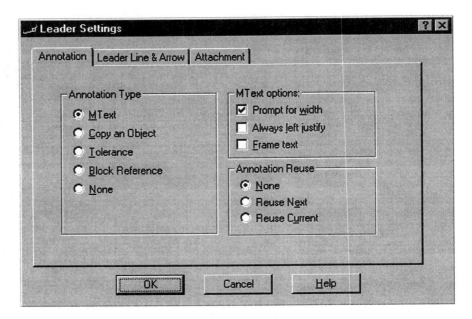

8. Click on the area of the Leader Settings dialog box designed to draw a box around the text next to the leader.

Level I Objective 7.02

9. You zoom into a portion of your drawing and save this display under a unique name. The name of this screen position is saved with the drawing file and can be retrieved for later use. What command is used to perform the task just described?

   (A) PAN
   (B) ZOOM
   (C) VIEW
   (D) WINDOW

Level I Objective 2.02

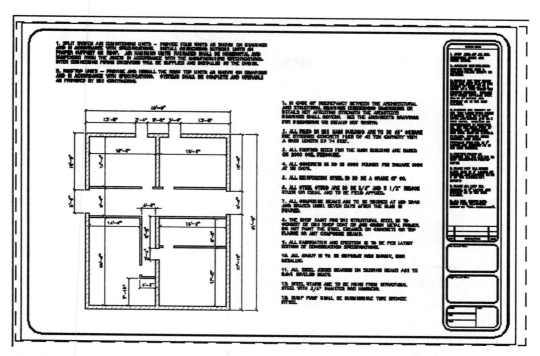

10. Open the drawing called BUILDING PLAN 3. After performing a search on all text, what is the name of misspelled word?

     (A) BULDING
     (B) MAXMUM
     (C) NEX
     (D) RECIEVED

Level I Objective 6.04

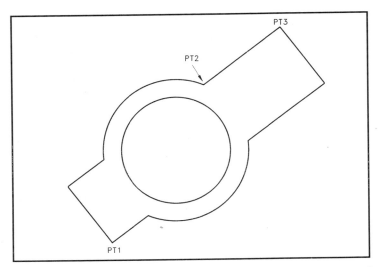

11. Open the drawing SCALE3. Reduce the size of all line, arc, and circle objects using the endpoints at PT2 and PT3 as the reference edge. Use the endpoint at PT1 as the base point for the scale operation. When prompted, specify a new length of 2. When finished, what is the total area of the shape minus the hole?

    (A) 7.1678
    (B) 7.1682
    (C) 7.1686
    (D) 7.1690

Level I Objective 5.09

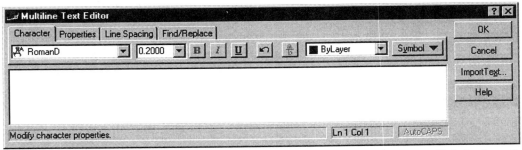

12. In the illustration above of the Multiline Text Editor dialog box, click in the tab that holds various justification modes.

Level I Objective 6.03

13. You perform a zoom operation that zooms to display the entire drawing based on the drawing limits. Which option of the ZOOM command performs this task?

   (A) All
   (B) Extents
   (C) Previous
   (D) Window

Level I Objective 2.01

14. What does the "TTR" option of the CIRCLE command allow you to construct?

   (A) Traced circle
   (B) Tangent circle
   (C) Trimmed circle
   (D) Transparent circle

Level I Objective 3.07

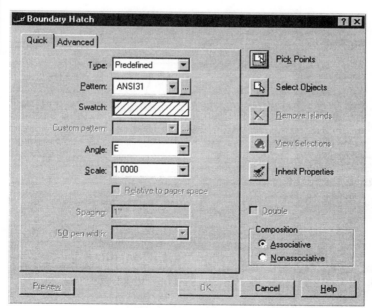

15. In the illustration above of the Boundary Hatch dialog box, the current hatch pattern is ANSI31. Click on an area of this dialog box to change to a different hatch pattern.

Level I Objective 6.05

16. You activate an Osnap mode which prompts you to enter an offset value. This Osnap mode is commonly used to reference a selected point on the drawing. Which Object Snap mode describes this task?

(A) Quick
(B) From
(C) Nearest
(D) Tracking

Level I Objective 3.11

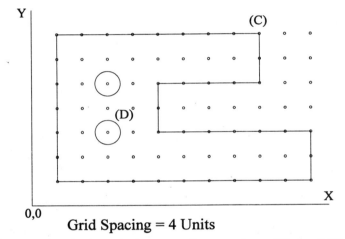

Grid Spacing = 4 Units

17. In the figure above, what is the relative coordinate value of Point "C" from the center of circle "D"?

(A) @-16.00.-24.00
(B) @-24.00,-16.00
(C) @16.00,24.00
(D) @24.00,16.00

Level I Objective 3.02

18. From the following list, what operation cannot be performed when using the Layer Properties Manager dialog box?

(A) Freezing a layer
(B) Creating a new layer
(C) Assigning color to a layer
(D) Changing objects to another layer

Level I Objective 1.01

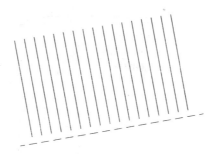

19. In the illustration above, the EXTEND command is used to extend all lines to the boundary edge identified by the hidden line. What would be the best selection mode to choose the objects to extend?

   (A) By a Fence
   (B) By a Window
   (C) By a Crossing box
   (D) To select the lines to extend individually

Level I Objective 5.01

20. At the "Command:" prompt, you click on an object. Grips appear at key locations on the object. You pick on one of the grips and it turns red by default. What are valid grip modes?

   (A) Move, Erase, Scale
   (B) Mirror, Scale, Fillet
   (C) Stretch, Rotate, Move
   (D) Chamfer, Stretch, Mirror

Level I Objective 5.15

21. You need to construct a line 20 units in length and in the 180 degree direction. You will be performing this operation using the Direct Distance mode of entry. You turn Ortho on, enter the LINE command and pick a point at the "Specify first point:" prompt. To construct the line at the "Specify next point or [Undo]:" prompt, you move your cursor to the left and type what value?

   (A) 20
   (B) -20,0
   (C) @-20,0
   (D) @20<180

Level I Objective 3.04

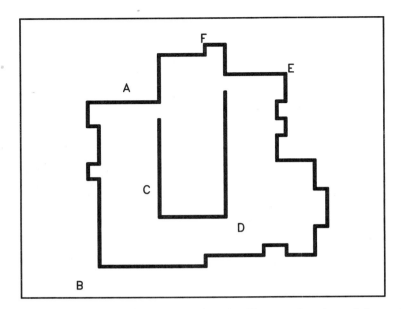

22. Open the existing drawing called FLOOR-1. Change the size of the walls using the STRETCH command according to the following specifications. Use a crossing box from "D" to "C" to stretch the walls straight up at a distance of 18'-0". Use a crossing box from "A" to "B" to stretch the walls directly to the left at a distance of 14'-0". What is the total area of the block wall including the three interior walls?

     (A) 304 sq. ft.
     (B) 307 sq. ft.
     (C) 310 sq. ft.
     (D) 313 sq. ft.

Level I Objective 5.10

23. You need to create an angled corner with the CHAMFER command. What rule must be followed when selecting the first and second objects before the chamfer is made?

     (A) Both objects must lie on different layers.
     (B) Both objects must not be parallel to each other.
     (C) Both objects must not be perpendicular to each other.
     (D) Both objects must be constructed with different linetypes.

Level I Objective 5.13

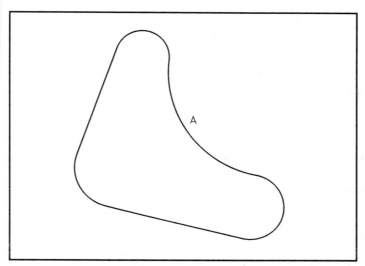

24. Open the existing drawing called BRACKET. Turn on layers Object and Dim. The perimeter consists of a single polyline object. Convert this single polyline object back into individual line and arc segments. What is the total length of arc "A"?

    (A) 6.013
    (B) 6.016
    (C) 6.019
    (D) 6.022

Level I Objective 1.02

25. You have just placed a hatch pattern in your drawing; however, you notice that the lines that make up the hatch pattern are closely spaced together. The spacing between these lines needs to be increased. What command is used to perform this task?

    (A) ERASE
    (B) EXPLODE
    (C) HATCHEDIT
    (D) SCALE

Level I Objective 6.06

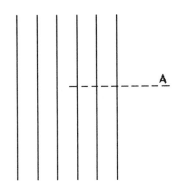

26. You enter the TRIM command. In the illustration above, Line "A" is selected as the cutting edge. You activate the Extend Edge option. How many lines will be trimmed to cutting edge "A"?

   (A) 3 lines
   (B) 4 lines
   (C) 5 lines
   (D) 6 lines

Level I Objective 5.12

27. You have constructed a drawing in real world units. You want to plot the drawing at a scale of 1/2 of its original size. What value is used in the Plot Scale area of the Plot dialog box if plotting from Model Space?

   (A) 1
   (B) 1=0.50
   (C) 0.50=1
   (D) 0.50

Level I Objective 9.02

28. What would you enter at the "Select objects" prompt to retrieve a selection set that had just been created?

   (A) All
   (B) Last
   (C) Previous
   (D) Crossing Polygon

Level I Objective 5.02

29. Which of the following statements is true when constructing 3 Point arcs with the ARC command?

(A) They must have a center identified.

(B) 3 Point arcs can only be constructed in the clockwise direction.

(C) 3 Point arcs must be constructed in the counterclockwise direction.

(D) 3 Point arcs can be drawn in either clockwise or counterclockwise directions.

Level I Objective 3.08

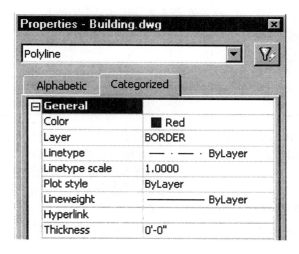

30. You have selected a polyline in a drawing. You activate the Properties window illustrated above. All color assignments need to be controlled by layers. How do you change the color of the polyline from Red to Bylayer in the Properties window?

(A) The color property must be changed through the CHPROP command.

(B) Click on the color red in the Properties window and choose the correct color from the drop down list.

(C) Dismiss the Properties window and change the color in the Layer Properties Manager dialog box.

(D) The Properties dialog box only allows you to view the properties of an object. Use the CHANGE command and the Properties option to change the color.

Level I Objective 5.17

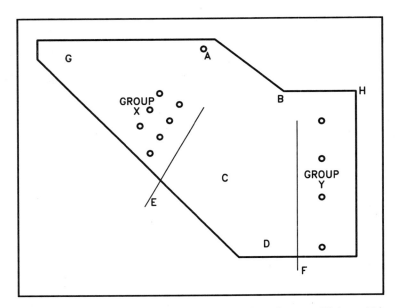

31. Open the existing drawing called GUSSET. Turn layers Object, Dim, and Gage On. Array hole "A" in a rectangular pattern to the left with 1 row and 9 columns. Use a spacing of 3 units between each column. Stretch vertex "B" 4 1/2 units to the left. Copy and flip the 7 holes in "Group X" using line "E" as the mirror line. Copy and flip the 4 holes in "Group Y" using line "F" as the mirror line. What is the area of the Gusset with all holes removed?

    (A) 1028.88
    (B) 1028.92
    (C) 1028.96
    (D) 1029.00

    Level I Objective 5.07

32. You need to create a symbol of a threaded screw. This symbol needs to be inserted into other drawing files. What command is used to create a global symbol?

    (A) BLOCK
    (B) BMAKE
    (C) SYMBOL
    (D) WBLOCK

    Level I Objective 8.01

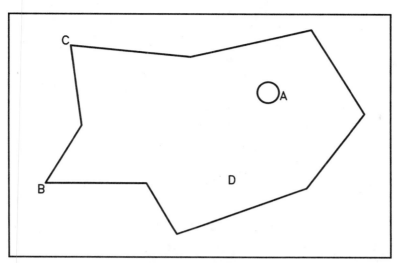

33. Open the existing drawing called BOUNDARY. Array hole "A" in a rectangular pattern of 3 rows and 3 columns. The spacing between rows is 1.75 units
down and the spacing between columns is 2.125 units to the left. What is the
X-Y coordinate value of the center of hole "D"?

    (A) 12.13,5.10
    (B) 12.17,5.10
    (C) 12.17,5.14
    (D) 12.21,5.14

Level I Objective 5.06

34. Enter AutoCAD and create a new drawing. Set the units to 3 decimal place
accuracy. Draw 4 points in the following exact order:
    Point #1 at X = 0.750, Y = 7.125
    Point #2 at X = 3.750, Y = 2.375
    Point #3 at X = 7.000, Y = 4.500
    Point #4 at X = 11.125, Y = 1.125.

If these four points were connected in this order, what would the total length be?

    (A) 14.831
    (B) 14.837
    (C) 14.843
    (D) 14.850

Level I Objective 3.01

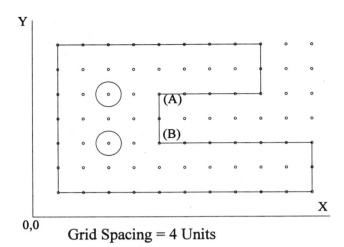

Grid Spacing = 4 Units

35. In the figure above, what is the polar coordinate value from Point "B" to Point "A"?

    (A) @8.00<360
    (B) @8.00<0
    (C) @8.00<90
    (D) @8.00<180

Level I Objective 3.03

36. To remove unused items in the drawing database, which compresses the current drawing, the PURGE command is used. From the following list, what items cannot be purged from a drawing?

    (A) Blocks
    (B) Layers
    (C) Text styles
    (D) Views

Level I Objective 8.04

37. You use the BREAK command to remove part of an object. When removing a segment of a circle using the BREAK command, in which direction will the break occur?

    (A) Clockwise
    (B) Counterclockwise
    (C) Horizontal
    (D) Vertical

Level I Objective 5.14

Create a new drawing of Plate1 illustrated below. Use the LIMITS command and set the upper right corner of the screen area to a value of 36.000,24.000. Use the Units Control dialog box set to decimal units. Set the number of digits to the right of the decimal point from 4 to 3. Accept the defaults for the remaining prompts. Begin this drawing by placing the center of the 4.000 diameter circle at absolute coordinate (16.000,13.000). When finished, answer the following question below:

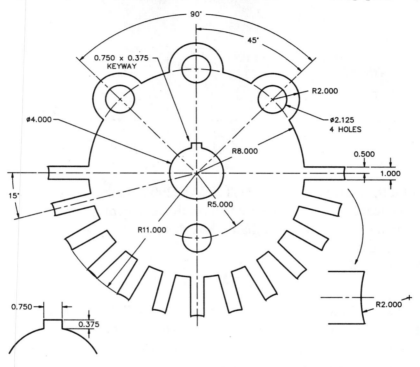

38. What is the total area of Plate1 with all holes including the hole with keyway removed?

     (A) 232.259
     (B) 232.265
     (C) 232.271
     (D) 232.277

Level I Objective 4.01

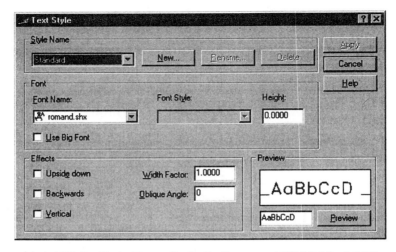

AUTOCAD 2000   AUTOCAD 2000
A   B

39. By default, entering the text "AUTOCAD 2000" in the illustration above displays the text at "A". You desire to have the text display similar to "B". Click in the area of the Text Style dialog box designed to make a change and condense the text.

Level I Objective 6.01

40. You have the ability to open up numerous AutoCAD drawings at the same time. This enables you to copy and paste objects from one drawing into another drawing very efficiently. You can also copy the properties of an object from one drawing to another using the Match Properties tool. What do all of the above tasks describe.

    (A) Expanded Reach
    (B) Streamlined Output
    (C) Extended AutoCAD
    (D) Multiple Drawing Environment

Level I Objective 8.02

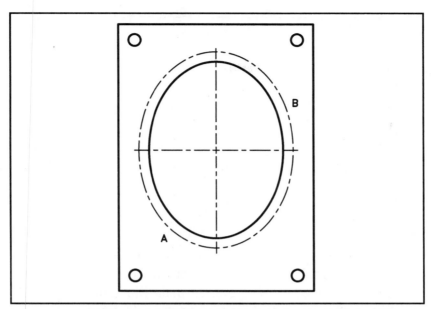

41. Open the existing drawing called SPEAKER. A block called "CIRCLE" exists in the database of this drawing. Place this block called "CIRCLE" equally spaced 12 times along the elliptical center line. What is the distance from the center of hole "A" to the center of hole "B"?

    (A) 10.567
    (B) 10.568
    (C) 10.569
    (D) 10.570

Level I Objective 3.10

42. A special tool exists that allows you to drag and drop content such as blocks, layers, layouts, and linetypes from a browser into any drawing. What is the name of this tool just described?

    (A) Windows Explorer
    (B) Extended AutoCAD
    (C) AutoCAD DesignCenter
    (D) The INSERT command

Level I Objective 8.03

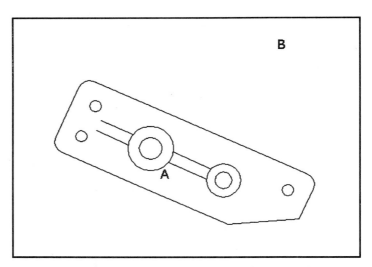

43. Open the drawing file called VIEW. This drawing was mistakenly rotated at an unknown angle. It is displayed in this angle in the illustration above. Rotate the entire drawing so the longest side is at an angle of 43 degrees, using the center of circle "A" as the center of the rotation. When finished, what is the new X,Y coordinate value of the intersection at "B"?

   (A) 13.9718, 14.5116
   (B) 13.9728,14.5116
   (C) 13.9728,14.5126
   (D) 13.9738,14.5126

Level I Objective 5.08

44. Of the following objects, which cannot be extended using the EXTEND command?

   (A) Arcs
   (B) Lines
   (C) Rays
   (D) Xlines

Level I Objective 5.11

45. You have several line segments in a drawing connected together. One of the segments is a polyline. You need to connect all line segments together to form a single polyline object. This can easily be performed using which of the following options of the PEDIT command?

   (A) Join
   (B) Block
   (C) Attach
   (D) Connect

Level I Objective 5.16

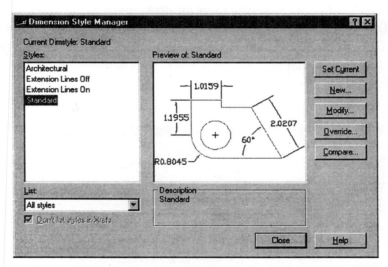

46. Click on the area of the Dimension Style Manager dialog box designed to change to a different dimension style.

Level I Objective 7.03

47. You place a series of notes in your drawing. Numerous lines of text make up the notes. All lines of text are a single object type. What command was used to create this object type?

   (A) DTEXT
   (B) MTEXT
   (C) TEXT
   (D) TXT

Level I Objective 6.02

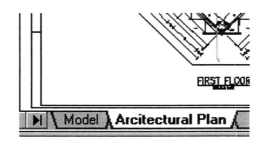

48. The layout name Arcitectural Plan is spelled incorrectly. How do you rename this layout?

(A) Single click on the layout to bring up a cursor menu and pick Rename.

(B) Right click on the layout to bring up a cursor menu and pick Rename.

(C) Double click on the layout to bring up a cursor menu and pick Rename.

(D) Double click while pressing the CTRL key to bring up a cursor menu and pick Rename.

Level I Objective 9.03

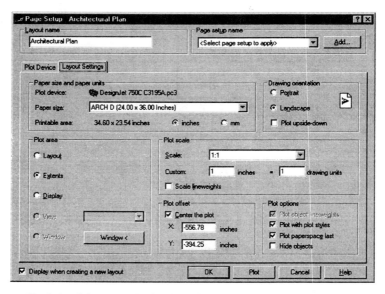

49. A layout called Architectural Plan has been created. You make various changes to the Paper size, Plot scale, Plot offset and Plot area. Click on the area of the Page Setup dialog box illustrated above designed to save these settings under a user defined name that can be retrieved for later use.

Level I Objective 9.01

Start a new drawing called Plan1. Change from decimal units to architectural units. Keep all remaining default values. **All wall thicknesses measure 4"**. Answer the question below regarding this drawing.

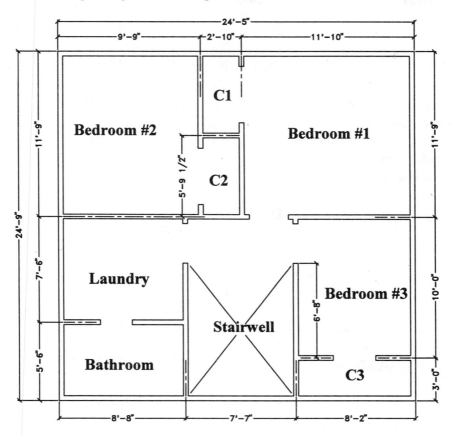

50. What is the total area in square feet of all bedrooms; closets C1, C2, and C3; the laundry; and bathroom?

    (A) 448 sq. ft.
    (B) 451 sq. ft.
    (C) 454 sq. ft.
    (D) 457 sq. ft.

Level I Objective 5.05

Answers to the AutoCAD 2000 Level I Exit Exam

1. C

2. B

3. B

4. In the Polar Angle measurement area, click in the radio button adjacent to Relative to last segment.

5. D

6. B

7. A

8. Under the MText options: area, click in the box adjacent to Frame text.

9. C

10. B

11. A

12. Click on the Properties tab of the Multiline Text Editor dialog box.

13. A

14. B

15. Click in the box adjacent to Pattern: or click on the small box that holds the three dots [...]

16. B

17. D

18. D

19. A

20. C

21. A

22. C

23. B

24. B

25. C

26. D

27. C

28. C

29. D

30. B

31. C

32. D

33. B

34. A

35. C

36. D

37. B

38. B

39. In the Effects area of the Text Style dialog box, click in the box adjacent to Width Factor.

40. D

41. D

42. C

43. B

44. D

45. A

46. Click on the Set Current button.

47. B

48. B

49. Click in the box below Page setup name.

50. B

Notes

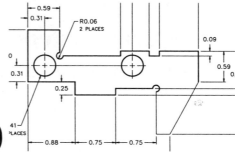

Chapter 6

AutoCAD 2000 Level II Exam Categories and Objectives

The AutoCAD 2000 Level II Assessment Exam consists of drawings and general knowledge questions that cover various advanced AutoCAD topics. Inquiry commands are used to analyze each drawing question. This takes the form of using such commands as AREA, DIST, ID, and LIST for performing various calculations on each drawing. Knowledge of using the Properties dialog box and the ability to create selection sets using Quick Select would also be helpful.

General knowledge questions may take the form of the following question types:

> Single answer multiple choice
> Hot spot

This chapter outlines the categories that make up the AutoCAD 2000 Level II Exam complete with the number of questions and a topic percentage that relates to the entire exam. Each category is further outlined with a detailed listing of the objectives an individual must master to be successful with passing the AutoCAD 2000 Level II Exam.

AutoCAD 2000
Level II Exam Categories

The AutoCAD 2000 Level II Exam consists of single answer multiple choice
questions and hot spot questions. Use the chart below for a breakdown on the
question categories, the number of questions per category, and the weight they carry
in the AutoCAD 2000 Level II Exam.

AutoCAD 2000 Level II Exam Categories	Number of Questions	Percentage of Overall Score
1 Drawing Objects	4	8%
2 Editing	13	26%
3 Dimensioning	2	4%
4 Managing Content	3	6%
5 Plotting	9	18%
6 Object, Layer, and Point Control	3	6%
7 Attributes, External References and Images	9	18%
8 Advanced Preference Settings and System Variables	3	6%
9 Customization	4	8%
Total	50	**100%**

AutoCAD 2000
Level II Exam Objectives

Category 1
Drawing Objects
Obj. 1.01   Draw objects using Relative Coordinates
Obj. 1.02   Draw objects using Polar Coordinates
Obj. 1.03   Use the MEASURE and DIVIDE commands
Obj. 1.04   Know the meaning of the @ symbol

Category 2
Editing
Obj. 2.01   Use the ARRAY command
Obj. 2.02   Use the MIRROR command
Obj. 2.03   Use the ROTATE command
Obj. 2.04   Use the SCALE command
Obj. 2.05   Use the STRETCH command
Obj. 2.06   Use the LENGTHEN command
Obj. 2.07   Use the TRIM command
Obj. 2.08   Use the FILLET and CHAMFER commands
Obj. 2.09   Group objects
Obj. 2.10   Use PEDIT to convert and join objects into a pline
Obj. 2.11   Edit a Hatch Pattern
Obj. 2.12   Add and subtract areas using the AREA command
Obj. 2.13   Apply linear or polar snaps using grips with the SHIFT key

Category 3
Dimensioning
Obj. 3.01   **Use QDIM***
Obj. 3.02   Use the Dimension Style Manager

*Note: Objectives in bold apply only to the AutoCAD 2000 exam and NOT the
          AutoCAD LT 2000 exam

Category 4
Managing Content
Obj. 4.01   View and copy content with DesignCenter
Obj. 4.02   Transfer data using the Multiple Drawing Environment
Obj. 4.03   Perform searches for content using AutoCAD DesignCenter

Category 5
Plotting
Obj. 5.01   Design a layout
Obj. 5.02   Control layer visibility by viewport
Obj. 5.03   Use PSLTSCALE to control linetype scaling in different floating viewports
Obj. 5.04   Scale dimensions in different floating viewports using a DIMSCALE of 0
Obj. 5.05   Know the purpose and function of locking viewports
Obj. 5.06   **Know how to clip existing viewports***
Obj. 5.07   Create and use Plot styles
Obj. 5.08   Know how to create and view .DWF files
Obj. 5.09   **Know how to use the Batch Plot Utility to plot multiple drawings***

Category 6
Object, Layer and Point Control
Obj. 6.01   **Create a selection set using the FILTER command***
Obj. 6.02   **Use the Extension and Parallel object snaps***
Obj. 6.03   Apply filters to layers

*Note: Objectives in bold apply only to the AutoCAD 2000 exam and NOT the AutoCAD LT 2000 exam

Category 7
Attibutes, External References and Images

Obj. 7.01   Understand the purpose and use of elements of attribute definitions

Obj. 7.02   Use the External Reference Manager

Obj. 7.03   Use the OLE scaling capability

Obj. 7.04   **Edit external references in-place***

Obj. 7.05   Extract attribute information using DDATTEXT

Obj. 7.06   Distinguish between attached and overlaid XREFs

Obj. 7.07   Identify the syntax of layer names of XREFs that are bound or attached

Obj. 7.08   Distinguish between bind and insert options in the Bind Xrefs dialog box

Obj. 7.09   Redefine attributes using the Properties window

Category 8
Advanced Preference Settings and System Variables

Obj. 8.01   Use the Dimension Edit menu

Obj. 8.02   Know the functions of the Plotting tab of the Options dialog box

Obj. 8.03   **Know the functions of the Profiles tab of the Options dialog box***

Category 9
Customization

Obj. 9.01   Create a new tool button macro

Obj. 9.02   Know the purpose of the menu files .mnu, .mns, .mnc, .mnl

Obj. 9.03   **Know how to load AutoLISP files***

Obj. 9.04   Identify the syntax for script files

*Note: Objectives in bold apply only to the AutoCAD 2000 exam and NOT the AutoCAD LT 2000 exam

Notes

Chapter 7

AutoCAD 2000 Level II Pretest

This pretest is designed to assess your current AutoCAD 2000 skill and knowledge levels. The AutoCAD 2000 Level II Pretest consists of 50 general knowledge questions mixed with performance-based drawing questions. There is no time limit to complete this assessment test. However, this test is designed to be completed in 2 hours or less, which would demonstrate use of the software in a productive manner. Question types include single-answer multiple-choice and hot-spot areas. Numerous questions have been designed around actual images an individual would be confronted with in the production drawing environment.

Two types of drawings are present in this pretest. Most of the drawings are already created up to a certain point. For these cases, open the drawing file and follow the steps that direct you to perform certain operations before attempting any of the questions that relate to the drawing. All drawings to open are provided on the disk supplied with this manual. Create a folder called \ASSESSMENT and load all drawing files there. Another type of drawing requires you to construct a new object from the image provided and answer the questions that follow the drawing to test your accuracy.

Work through this pretest at a good pace paying strict attention to the amount of time spent on each question. Answers for each Level II Pretest question are located at the end of this chapter.

Notes

Provide the best answer for each of the following questions.

1. If an externally referenced drawing called House.Dwg contains a layer called FND, what will the name of the layer in the current drawing be?
   (A) FND
   (B) HOUSE-FND
   (C) HOUSE|FND
   (D) HOUSE/FND

Level II Objective 7.07

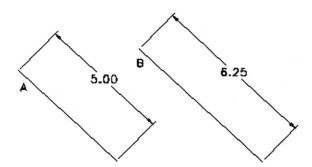

2. In the illustration above, the line segment at "A" with the 5.00 dimension needs to be extended to a new value of 6.25 in "B" without using a boundary edge. What command is used to accomplish this task?
   (A) CHANGE
   (B) EXTEND
   (C) LENGTHEN
   (D) MODIFY

Level II Objective 2.06

3. You need to create a layer called "Object" through a script file. From the following, identify the correct code used for performing this task?
   (A) -la object
   (B) -la new object
   (C) layer new object
   (D) layer object new

Level II Objective 9.04

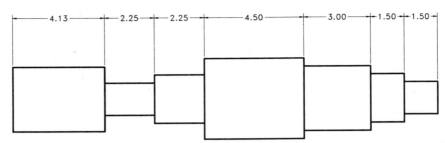

4. You use the QDIM command to dimension the object illustrated above. Which Quick Dimension mode is depicted in this illustration?
    (A) Baseline
    (B) Continuous
    (C) Ordinate
    (D) Staggered
Level II Objective 3.01

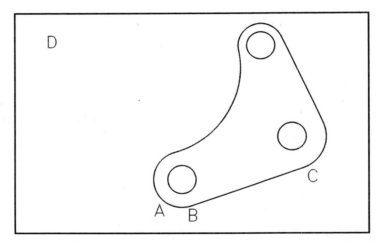

5. Open the drawing called ROTATE. Using the center of arc "A" as the base point of the rotation, rotate the object so that line "BC" has a new angle of 90 degrees. When finished, what is the XY coordinate value of the center of arc "D"?
    (A) 0.91,5.41
    (B) 0.91,5.44
    (C) 0.91,5.47
    (D) 0.91,5.50
Level II Objective 2.03

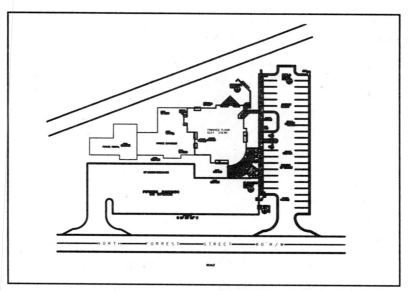

6. Open the existing drawing called COMPLEX EXPANSION. What is the total
number of layers assigned the color #15 in the database of this drawing?

    (A) 5
    (B) 7
    (C) 13
    (D) 19

Level II Objective 6.03

7. At the "Select objects:" prompt of the TRIM command, you press the ENTER
key. What occurs in the current drawing as a result of this action?

    (A) Nothing. You are returned to the Command: line.
    (B) All visible objects are considered cutting edges but do not highlight.
    (C) Nothing. The message "Invalid" occurs at the Command: line.
    (D) All objects in the drawing highlight because they are considered cutting
        edges.

Level II Objective 2.07

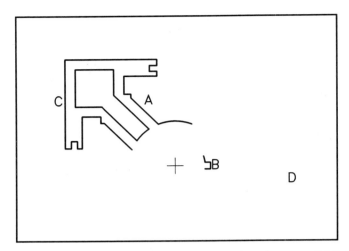

8. Open the existing drawing called FMS. Array the shape at "A" four times in a complete circle. Use the center mark as the center of the array. Then array the tooth at "B" 14 times in a complete circle using the center mark as the center of the array. Convert the outer profile, 4 inner shapes, and the 14 teeth into individual polyline objects. What is the surface area of the Fms extrusion with all four pockets and the tooth area removed?

   (A) 34.819
   (B) 34.824
   (C) 34.829
   (D) 34.834

Level II Objective 2.01

9. You edit an AutoCAD template menu file. When the template menu has been compiled, an additional menu file is automatically generated. This ASCII file is usually referred to as the source menu file. What extension does this source menu file have?

   (A) .mnc
   (B) .mnl
   (C) .mns
   (D) .mnu

Level II Objective 9.02

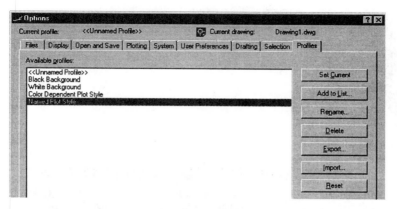

10. Click on the proper area of the Profiles tab of the Options dialog box designed
    to make the "Named Plot Style" profile active.

Level II Objective 8.03

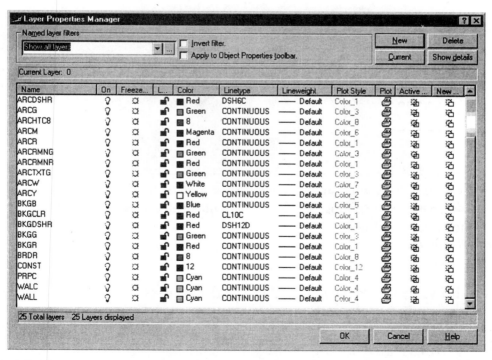

11. Click in the area of the Layer Properties Manager dialog box designed to freeze
    the layer called WALL only in the current viewport.

Level II Objective 5.02

12. What option of the PEDIT command is used to convert a series of line segments into a single polyline?

> (A) Close
> (B) Fit
> (C) Join
> (D) Spline

Level II Objective 2.10

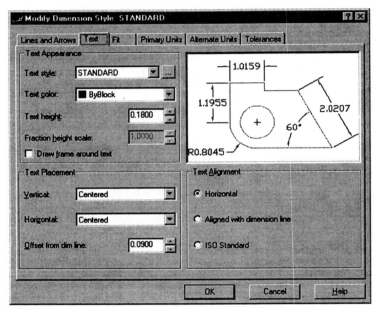

13. From the illustration above, click on the area of the Text tab of the Modify Dimension Style dialog box designed to make all dimension text parallel with the dimension line.

Level II Objective 3.02

14. What value should the PSLTSCALE variable be set to for linetype scaling to be controlled by the scale factor of a viewport in a layout?

> (A) 0
> (B) 1
> (C) 2
> (D) 3

Level II Objective 5.03

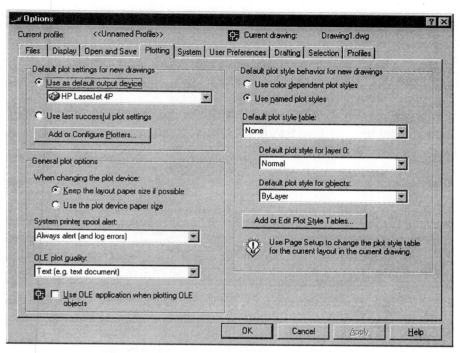

15. From the illustration above, click on the area of the Options dialog box designed to make changes to a named plot style table.
Level II Objective 8.02

16. Start a new drawing. Draw a rectangle and specify the first corner point at coordinate 4.1509,2.9173. Locate the other corner point 7.6283 units along the X axis and 2.1371 units along the Y axis from the last point. When finished, what is the total area of this rectangle?

    (A) 16.3019
    (B) 16.3024
    (C) 16.3029
    (D) 16.3034

Level II Objective 1.01

17. In the illustration at the right of the Properties window, an attribute is assigned with a tag name of "PRODUCT_NAME." When an attribute value is entered, it needs to be hidden on the display screen. Click on the area of this window designed to perform this task.

Level II Objective 7.09

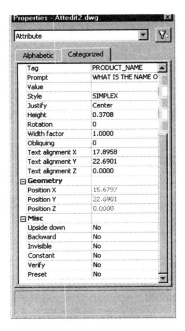

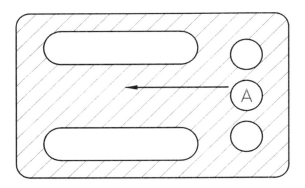

18. The object illustrated above was crosshatched using the BHATCH command. You need to move circle "A" to the left. What effect does this action have on the crosshatch pattern?

   (A) The hole moves but the hatch pattern remains unchanged.

   (B) The hole moves and the hatch pattern updates to this change.

   (C) The hole cannot be moved once associated with a hatch pattern.

   (D) The hole moves and the hatch pattern transfers to the inside of the hole.

Level II Objective 2.11

Start a new drawing called Bldg1. Change from decimal units to architectural units. Keep all remaining default values. **All block wall thicknesses identified by crosshatching measure 12". All other interior wall thicknesses measure 6".** Do not add any dimensions to this drawing. When finished, answer the question at the bottom of this page.

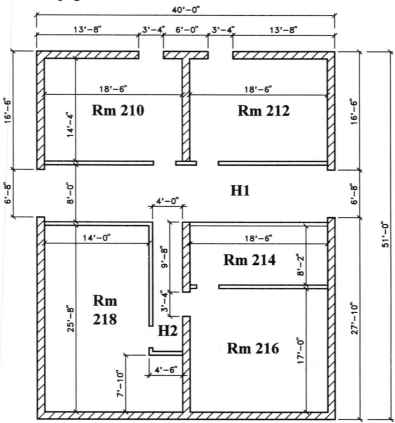

19. Convert all concrete block walls identified by the crosshatching pattern into individual polyline objects. This should create 5 polyline objects. What is the total area of all concrete block walls identified by the crosshatching pattern?

    (A) 192 sq. ft.
    (B) 195 sq. ft.
    (C) 198 sq. ft.
    (D) 201 sq. ft.

Level II Objective 2.12

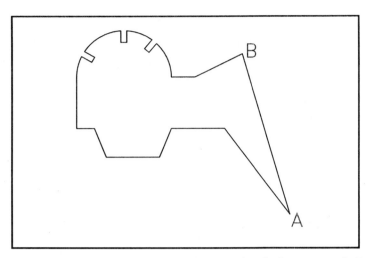

20. Open the drawing HITCH. Increase the length of Line AB so it has a new total
    length of 5.00 units. The remainder of the objects should also increase in
    proportion to this line. What is the new area?

    (A) 14.47
    (B) 14.51
    (C) 14.55
    (D) 14.59

Level II Objective 2.04

21. Start a new drawing. Construct an arc using the Start, Center, and End modes.
    For the starting point, enter a coordinate value of 4.29,2.56. For the arc center,
    enter 6.63,4.64. For the end of the arc, enter a value of 5 units at a 65 degree
    angle from the last known point. When finished, what is the length of the arc
    segment?

    (A) 11.1120
    (B) 11.1123
    (C) 11.1126
    (D) 11.1129

Level II Objective 1.02

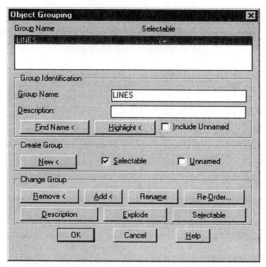

22. In Object Grouping dialog box illustrated above, a group called LINES is already created. Click on the area of the dialog box designed to delete the group name LINES from the dialog box.
Level II Objective 2.09

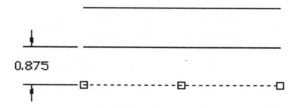

23. In the illustration above, a line is selected and grips appear. The middle grip is then picked. Multiple Grip Move/Copy mode is invoked and the line is copied 0.875 units. What key activates the grid snap which enables the other lines to be copied by picking new screen locations at 0.875 increments?
    (A) ALT
    (B) CTRL
    (C) ESC
    (D) SHIFT
Level II Objective 2.13

24. Drawing files can be viewed over the Internet. These files must be converted into a special format while inside of AutoCAD. What extension identifies the special file format required for viewing drawings over the Internet?

(A) DWF
(B) DWK
(C) DXF
(D) HTM

Level II Objective 5.08

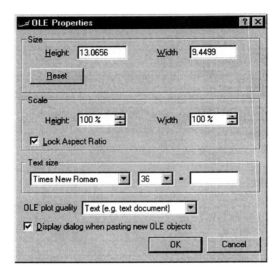

25. You embed a document from Microsoft Word into your AutoCAD drawing. You notice that document is hard to read and needs to be scaled larger in size. What action will activate the OLE Properties dialog box illustrated designed to change the scale of the Word document in the AutoCAD drawing?

(A) Double click with the left mouse button on the Word document.
(B) Single click on the document while pressing down the SHIFT key.
(C) Right click on the document and pick Properties... from the cursor menu.
(D) Single click on the document and pick Properties... from the cursor menu.

Level II Objective 7.03

26. Illustrated at the right is a list of items that have been generated through the ATTEXT command. By what format is this list organized?

    (A) CDF
    (B) DXF
    (C) DXX
    (D) SDF

Level II Objective 7.05

'CPU','DELL', 2400.00
'MONITOR','DELL', 400.00
'MONITOR','DELL', 400.00
'CPU','DELL', 2400.00
'CPU','DELL', 2400.00
'MONITOR','DELL', 400.00
'CPU','DELL', 2400.00
'MONITOR','DELL', 400.00

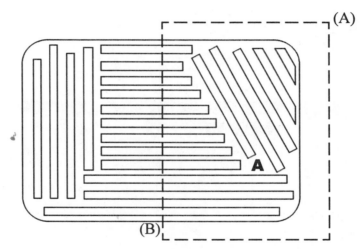

27. Open the existing drawing called SHIELD illustrated above. Numerous rectangular shapes have been constructed inside of another rectangular shape that has its corners filleted. All rectangular shapes are made up of individual objects. Stretch the Shield 1.375 units to the right using a crossing box from "A" to "B". When finished, what is the calculated area of "A" of the Shield?

    (A) 48.735
    (B) 48.739
    (C) 48.743
    (D) 48.747

Level II Objective 2.05

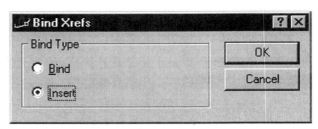

28. You have externally referenced the file "PIPE_FLOW" into your current drawing file. A block called "GLOBE" is brought over with this external reference file. Clicking on the Insert radio button of the Bind Xrefs dialog box illustrated above results in the block having what name?

    (A) GLOBE
    (B) PIPE_FLOW|GLOBE
    (C) PIPE_FLOW0GLOBE
    (D) PIPE_FLOW%_%GLOBE

Level II Objective 7.08

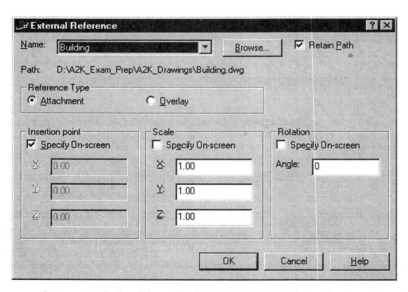

29. You are ready to attach the file called "BUILDING" into the host drawing as an external reference. However, you do not want other users to see "BUILDING" when they externally reference the host drawing into their drawing. Click on the area of the External Reference dialog box designed to perform this task.

Level II Objective 7.06

30. You need to plot numerous drawings in one plot operation. During the plot of these drawings, you have the capabilities of controlling layers, plot area, and space. What do these plotting features describe?

    (A) Batch Plot Utility

    (B) Print/Plot Configuration dialog box

    (C) The ability to plot using a script file

    (D) The ability to plot to the default system printing device

Level II Objective 5.09

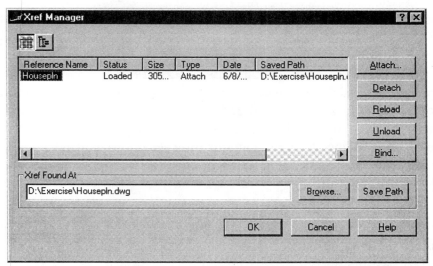

31. In the Xref Manager dialog box illustrated above, the external reference file HOUSEPLN needs to be made a permanent part of the current drawing database. What button is used to perform this operation?

    (A) Attach...

    (B) Bind...

    (C) Detach

    (D) Reload

Level II Objective 7.02

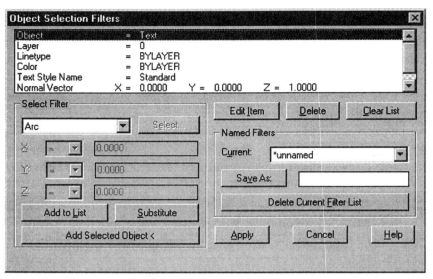

32. You have activated the Object Selection Filters dialog box illustrated above to create a selection set based on specific text objects in the current drawing. You forsee the need to use this filter selection set again. Click on the area of this dialog box designed to create a unique filter selection set name which can then be retrieved later on in the design process.
Level II Objective 6.01

33. From the following groupings, which item cannot be viewed or copied with the AutoCAD DesignCenter?
    (A) Layouts
    (B) Linetypes
    (C) Views
    (D) Xrefs
Level II Objective 4.01

34. How do you apply the Parallel Object Snap for constructing a line from an existing line segment?
    (A) Pick the existing line.
    (B) Pause over the existing line.
    (C) Right click on the existing line.
    (D) Hold down the SHIFT key and pick the existing line.
Level II Objective 6.02

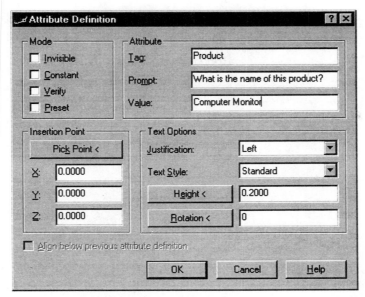

35. You are in the process of creating an attribute definition. In the illustration above, click in the area of the Attribute Definition dialog box designed to repeat the attribute prompt.

Level II Objective 7.01

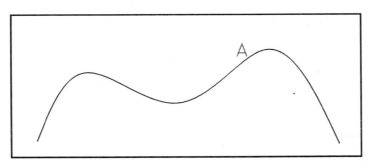

36. Open the drawing called SPLINE. Divide this object into 7 equal segments. When finished, what is the length of the segment between the two points identified as "A"?

    (A) 1.6197
    (B) 2.6202
    (C) 2.6207
    (D) 2.6212

Level II Objective 1.03

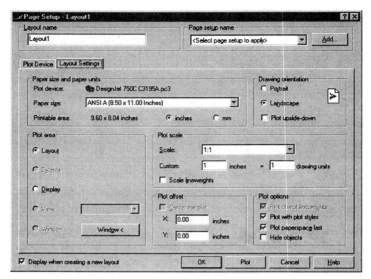

37. In the illustration above, click in the area of the Page Setup dialog designed to rename a layout.
Level II Objective 5.01

38. You have arranged two drawings on a single screen in a Multiple Drawing Environment. Both drawings can be seen at the same time. You select a series of objects in one drawing. You then drag these objects into the other drawing by holding down the left mouse button. What operation takes place as a result of performing this task?

    (A) Copy
    (B) Group
    (C) Insert
    (D) Move
Level II Objective 4.02

39. You enter a coordinate value of @4<180. What does the presence of the "@" symbol mean?

    (A) 0,0
    (B) origin
    (C) last point
    (D) next point
Level II Objective 1.04

40. Click on the button in the Viewports toolbar designed to convert an existing closed object into a viewport.

Level II Objective 5.06

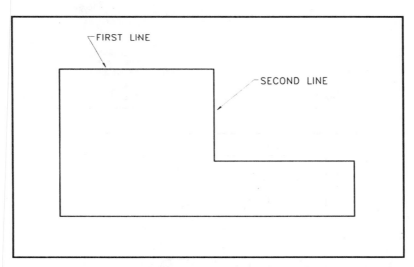

41. Open the drawing called CHAMFER. Set the chamfer length on the first line to 3.00 units. Also set the chamfer angle from the first line to 30 degrees. Create the chamfer using the illustration above. When finished, what is the area of the closed object?

    (A) 41.7635
    (B) 41.7637
    (C) 41.7640
    (D) 41.7643

Level II Objective 2.08

42. Click on the area of the Dimension Edit menu in the illustration at the right designed to change the number of decimal places of a dimension from 4 to 2.

Level II Objective 8.01

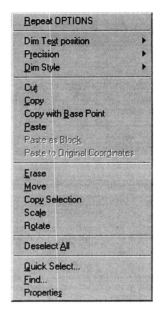

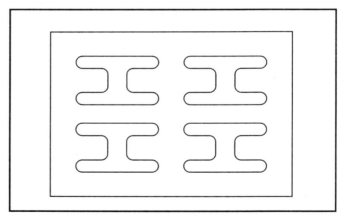

43. Open the drawing called STAMPING. Create mirrored images to form the part in the illustration above. Use points "A" and "B" as points for the mirror operation. When finished, what is the area of the plate with the four stamping shapes removed?

(A) 48.26
(B) 48.30
(C) 48.34
(D) 48.38

Level II Objective 2.02

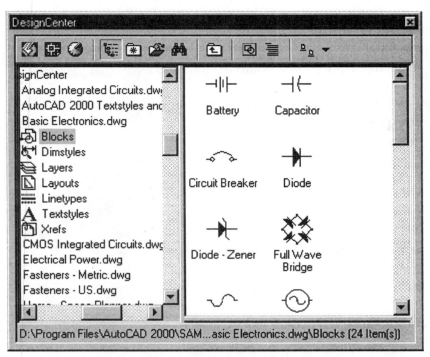

44. From the illustration provided, click on the area of the DesignCenter dialog box designed to take you to area to search for drawing files.
Level II Objective 4.03

45. You just created a plot style that is based on the object's color. Up to 255 color-dependent plot styles are available to you. What extension is this plot style identified by?

    (A) .ctb
    (B) .pc2
    (C) .pcp
    (D) .stb
Level II Objective 5.07

46. While in a drawing layout, you select a viewport at the Command prompt and right click on the viewport. The cursor menu displays in the illustration at the right. Click on the proper area designed to prevent the scale of the viewport from being changed.

Level II Objective 5.05

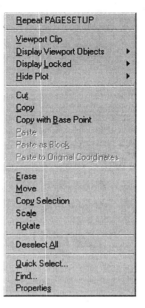

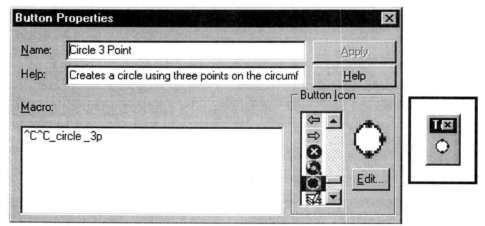

47. You create a toolbar that includes the button shown to the right in the illustration above. You need to modify the button icon. What action will display the Button Properties dialog box after issuing the TOOLBAR command?

    (A) Right click on the button.

    (B) Single click on the button.

    (C) Double click on the button.

    (D) Press the CTRL key while single clicking on the button.

Level II Objective 9.01

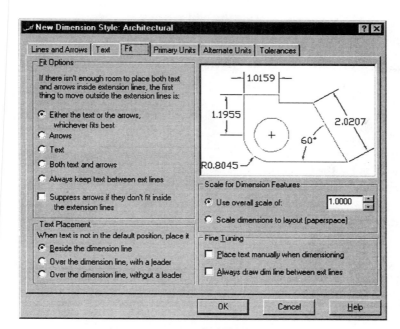

48. Click on the area of the New Dimension Style dialog box illustrated above designed to have the same effect as setting the system variable DIMSCALE to a value of 0 (zero).
Level II Objective 5.04

49. You select an external reference for in-place editing using the REFEDIT toolbar. After creating a working set of the individual objects to edit, what happens to the remainder of the objects that belong to the external reference?
   (A) They are locked and appear faded.
   (B) They are assigned to a layer and turned off.
   (C) They are assigned to a layer and made frozen.
   (D) Nothing. All objects appear the same on the display screen.
Level II Objective 7.04

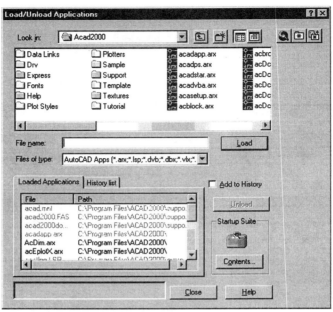

50. You need to load an AutoLISP file into your AutoCAD drawing. What command will display the Load/Unload Applications dialog box illustrated above?

    (A) APPLOAD

    (B) LOAD

    (C) LOADAPP

    (D) MENULOAD

Level II Objective 9.03

Answers to the AutoCAD 2000 Level II Pretest

1. C

2. C

3. C

4. B

5. B

6. A

7. B

8. B

9. C

10. Click on the Set Current button.

11. Click on the layer "WALLS". Then click on the sun icon located under the "Active VP freeze" column.

12. C

13. In the Text Alignment area, click on the radio button adjacent to "Aligned with dimension line".

14. B

15. Click on the Add or Edit Plot Style Tables... button in the Options dialog box.

16. B

17. In the Misc category of the Property window, change Invisible from No to Yes.

18. B

19. B

20. B

21. C

22. Click on the Explode button found under the Change Group area of the Object Grouping dialog box.

23. D

24. A

25. C

26. A

27. C

28. A

29. Click on the radio button adjacent to "Overlay" under the Reference Type area of the External Reference dialog box.

30. A

31. B

32. Click on the Save As: button under the Named Filters area of the Object Selection Filters dialog box.

33. C

34. B

35. Place a check in the box adjacent to "Verify" under the Mode area of the Attribute Definition dialog box.

36. B

37. Click in the Layout name box of the Page Setup dialog box.

38. A

39. C

40. Click on this button:

41. D

42. Click on the Precision area of the Dimension Edit menu.

43. C

44. Click on the binocular button at the top of the DesignCenter dialog box.

45. A

46. Click in the Display Locked area of the cursor menu

47. A

48. Click in the radio button adjacent to "Scale dimensions to layout (paperspace)".

49. A

50. A

Chapter 8

AutoCAD 2000 Level II Practice Test

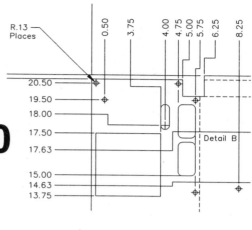

This AutoCAD 2000 Level II Practice Test has been developed for more practice in preparing for the Level II Assessment Exam. This practice test consists of 50 drawing and general knowledge questions. Question types include single-answer multiple-choice and hot-spot areas. Numerous questions have been designed around actual images an individual would be confronted with in the production drawing environment. There is no time limit to complete this assessment test. However, this test is designed to be completed in 2 hours or less, which would demonstrate use of the software in a productive manner.

Two types of drawings are present in this practice test. Most of the drawings are already created up to a certain point. For these cases, open the drawing file and follow the steps that direct you to perform certain operations before attempting to answer any of the questions that relate to the drawing. All drawings to open are provided on the disk supplied with this manual. Create a folder called \ASSESS-MENT and load all drawing files there. Another type of drawing requires you to construct a new object from the image provided and answer the question that follows the drawing to test your accuracy.

Work through this Practice Test at a good pace, paying strict attention to the amount of time spent on each question. Answers for all Level II Practice Test questions are located at the end of this chapter.

Notes

Provide the best answer for each of the following questions.

1. You create a circle. You now want to create a second circle and use the same center point as the first circle. What can you type to locate the last point without the aid of object snap modes or construction lines?

    (A) @
    (B) #
    (C) ^
    (D) <

Level II Objective 1.04

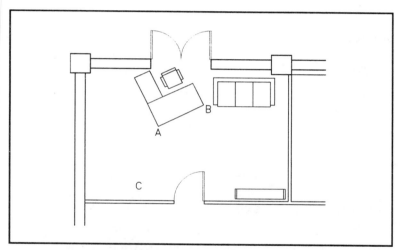

2. Open the drawing OFFICE PLAN. The desk and chair need to be rotated so line "AB" lies in the 180 degree position. Use the intersection at "A" as the base point of the rotation operation. When finished, what is the XY coordinate value of the corner of the desk at endpoint "C"?

    (A) 26'-0",68'-0"
    (B) 26'-1",68'-1"
    (C) 26'-1",68'-2"
    (D) 26'-2",68'-3"

Level II Objective 2.03

3. You design a script file that contains the following layer operation:

> **LAYER N Object**

The words "LAYER" and "Object" and the letter "N" are all separated by spaces. What does the presence of a space mean in a script file?
(A) Pause
(B) Enter
(C) No Meaning
(D) User Input
Level II Objective 9.04

4. What command allows you to scale a linetype in paper space based on the scale of the viewport in which it is visible?
(A) LTSCALE
(B) LTYPE
(C) PSLTSCALE
(D) PSPACE
Level II Objective 5.03

5. Start a new drawing. Draw a continuous polyline using the following information: Begin the polyline at coordinate 3.27,5.58. Draw the polyline segment 1.12 units along the X axis and 4.38 units along the Y axis from the last point. Draw the next polyline segment 6.38 units along the X axis and -3.67 units along the Y axis from the last point. Draw the last polyline segment 5.32 units along the X axis and 5.07 units along the Y axis from the last point. When finished, what is the total length of the polyline object?
(A) 19.08
(B) 19.13
(C) 19.18
(D) 19.23
Level II Objective 1.01

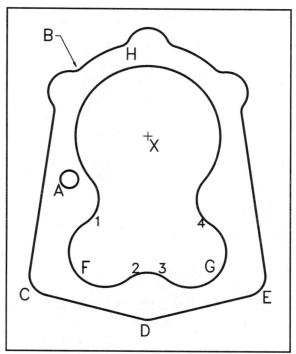

6. Open the existing drawing called SEAL. Array circle "A" five times to fill 240
   degrees in the clockwise direction using the intersection at "X" as the center of
   the array. Change all fillets of radius 0.250 units (similar to the radius at "B")
   to 0.375 radius. Using the center of the circle as the base point, copy circle "A"
   to the centers of arcs "C," "D," and "E." Reconstruct arcs "F" and "G" by using
   a new radius value of 0.800 for each. What is the area of the Seal with all circles
   and inner shape removed?

   (A) 11.649
   (B) 11.654
   (C) 11.659
   (D) 11.674

Level II Objective 2.01

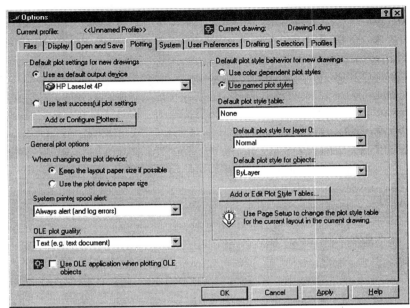

7. From the illustration above, click on the area of the Options dialog box designed to apply a plot style that is controlled by color to a new drawing.
Level II Objective 8.02

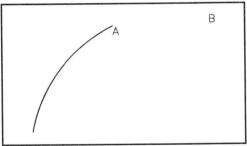

8. Open the drawing ARC. In the illustration above, extend the end of the arc at "A" so that the total length of the arc measures 11.75 units. When finished, are the XY coordinates of the end of the arc at "B"?

    (A) 13.31,7.89
    (B) 13.34,7.93
    (C) 13.34,7.97
    (D) 13.46,8.01

Level II Objective 2.06

9. What does a layout allow for?
   (A) arranging views with TILEMODE turned On
   (B) constructing a drawing at full size or real world units
   (C) creating of viewports using the VPORTS command
   (D) arranging and plotting of drawing details at different scales on the same drawing sheet

Level II Objective 5.01

10. Initially when beginning to customize a menu, you start out by making additions to a template menu file supplied when you load AutoCAD. What extension does this template menu file have?
    (A) .mnc
    (B) .mnl
    (C) .mnp
    (D) .mnu

Level II Objective 9.02

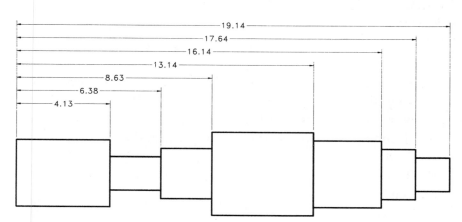

11. You use the QDIM command to dimension the object illustrated above. Which Quick Dimension mode is depicted in this illustration?
    (A) Baseline
    (B) Continuous
    (C) Ordinate
    (D) Staggered

Level II Objective 3.01

12. Several objects have been grouped under the name of "BOLTS". With Group mode turned on, all objects that belong to the group select when picking a single group member. How do you toggle Group mode off?

(A) CTRL-A
(B) CTRL-G
(C) CTRL-X
(D) F11

Level II Objective 2.09

13. Open a drawing called POLYGON. Construct a 6 sided polygon circumscribed about a circle. Use coordinate 5.00,4.00 as the center of the polygon. For the radius of the circle, enter a value 3.50 units at an angle of 135 degrees from the last known point. When finished, what are the XY coordinate value of the intersection at "A"?

(A) 8.87,2.93
(B) 8.90,2.95
(C) 8.90,2.98
(D) 8.94,2.98

Level II Objective 1.02

14. You have attached numerous drawing files as external references into the current drawing file. You are finished utilizing one of the external references and need to permanently remove it from the display screen. What option of the External References dialog box is used to perform this task?

(A) Bind
(B) Detach
(C) Reload
(D) Unload

Level II Objective 7.02

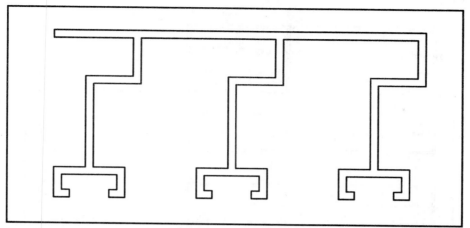

15. Open the drawing called MULTIPLE CHAMFER. Chamfer every corner of the object in the illustration above. Use equal chamfer distances of 0.15. When finished, what is the surface area of the object?
    (A) 52.8435
    (B) 52.8439
    (C) 52.8443
    (D) 52.8447

Level II Objective 2.08

16. Begin a new drawing. Construct a circle of radius 3.50 units. Measure increments of 1.70 units along the perimeter of circle. At what angle on the circle does the measuring begin?
    (A) 0
    (B) 90
    (C) 180
    (D) 270

Level II Objective 1.03

17. How do you activate the Extension Object Snap Mode on a line or arc segment?
    (A) Pick the endpoint of the object.
    (B) Pause over the endpoint of the object.
    (C) Right click on the endpoint of the object.
    (D) Hold down the CTRL key and pick the existing object.

Level II Objective 6.02

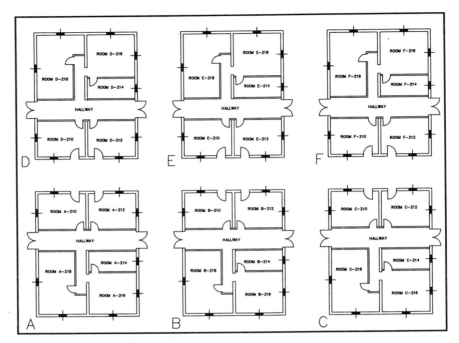

18. Open the existing drawing called BUILDINGS. What is the total number of layers assigned the color red and frozen in the database of this drawing?

    (A) 2
    (B) 7
    (C) 11
    (D) 14

Level II Objective 6.03

19. You open the Options dialog box and you change settings in various tabs. You have the ability of saving these changes under a unique name which can be retrieved at any time. Under which tab is this feature available?

    (A) Files
    (B) Profiles
    (C) System
    (D) User Preferences

Level II Objective 8.03

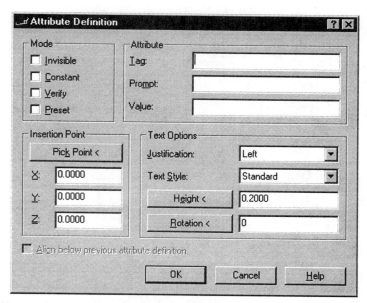

20. In the illustration above of the Attribute Definition dialog box, click on the proper mode designed to make the attribute value hidden on the display screen.
Level II Objective 7.01

21. You create the following tool button macro:

$$\wedge C \wedge C\_break \setminus f \setminus @$$

What is the purpose for the ^C^C?
    (A) Toggle the grid on or off.
    (B) Toggle Object Snap on or off.
    (C) Cancel the previous command.
    (D) Toggle running coordinate display on or off.
Level II Objective 9.01

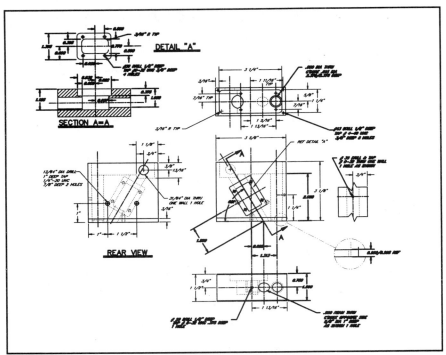

22. Open the existing drawing called BASE PLATE and enter the Paper space environment. What is the total number of text strings contained in the entire database of this drawing?

    (A) 32
    (B) 34
    (C) 36
    (D) 38

Level II Objective 6.01

23. You created a rectangular viewport. You now want to change this viewport to a different shape. Click on the button in the Viewports toolbar designed to clip an existing viewport into a different shape.

Level II Objective 5.06

CPU	DELL	2400.00
MONITOR	DELL	400.00
MONITOR	DELL	400.00
CPU	DELL	2400.00
CPU	DELL	2400.00
MONITOR	DELL	400.00

24. Illustrated above is a list of items that have been generated through the ATTEXT command. By what format is this list organized?

    (A) DXF
    (B) DXX
    (C) CDF
    (D) SDF

Level II Objective 7.05

25. You select an object and grips appear. You wish to move and copy the selected object. After creating the first copied object, what key do you hold down to create additional copies of the object at the same grip offset distance?

    (A) ALT
    (B) CTRL
    (C) ESC
    (D) SHIFT

Level II Objective 2.13

26. WIRES.DWG is a nested xref that is attached to another xref called ELECTRICAL.DWG. What mode of the External Reference dialog box was selected to display ELECTRICAL.DWG but ignore all objects that lie on WIRES.DWG?

    (A) Attach
    (B) Bind
    (C) Overlay
    (D) Reload

Level II Objective 7.06

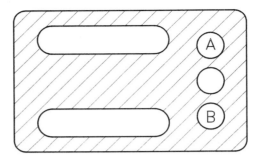

27. The object illustrated above was crosshatched using the BHATCH command. You need to erase circles "A" and "B". What effect does this action have on the crosshatch pattern?

    (A) The holes erase along with the hatch pattern.

    (B) The holes erase but the hatch pattern remains unchanged.

    (C) The holes erase and the hatch pattern updates to this change.

    (D) The holes cannot be erased once associated with a hatch pattern.

Level II Objective 2.11

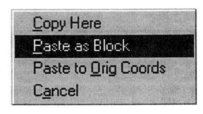

28. You have arranged two drawings on a single screen in a Multiple Drawing Environment. Both drawings can be seen at the same time. You select a series of objects in one drawing. You then drag these objects into the other drawing and notice the cursor menu appearing similar to the illustration above. What action displays this special cursor menu when dragging and dropping objects from one drawing into another?

    (A) Pressing and dragging the objects with the left mouse button.

    (B) Pressing and dragging the objects with the right mouse button.

    (C) Holding down the SHIFT key while pressing and dragging the objects with the left mouse button.

    (D) Holding down the CTRL key while pressing and dragging the objects with the right mouse button.

Level II Objective 4.02

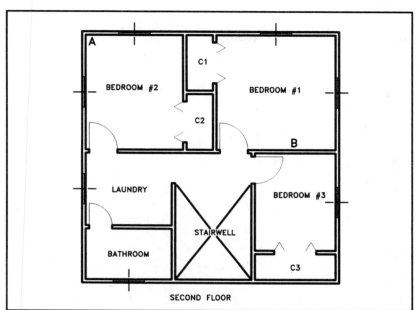

29. Open the existing drawing named PLAN. Stretch the right-most wall a distance of 5'-6" to the right. What is the total area of bedrooms #1, #2, and #3 closest to?

    (A) 374 sq. ft.
    (B) 377 sq. ft.
    (C) 381 sq. ft.
    (D) 385 sq. ft.

Level II Objective 2.05

30. You create a plot style that can be assigned to any object, regardless of the object's color. By what extension is this plot style identified?

    (A) .ctb
    (B) .pc2
    (C) .pcp
    (D) .stb

Level II Objective 5.07

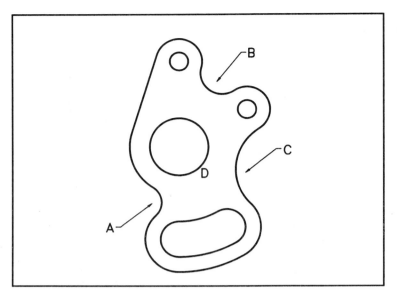

31. Open the existing drawing called GASKET. Reconstruct the following arcs
    using new radius values: Change radius "A" to a new value of 24 units; Change
    radius "B" to a new value of 25 units; Change radius "C" to a new value of 50
    units (all arcs must be tangent after these changes are made.) Convert the outer
    perimeter, inner slot, and three holes of the Gasket to individual polylines.
    What is the total area of the Gasket with all holes and slot removed?

    (A) 7224
    (B) 7228
    (C) 7232
    (D) 7236

Level II Objective 2.10

32. While in Layout mode, what command is used to make a layer visible in one
    viewport but frozen in all other viewports?

    (A) -LAYER
    (B) MS
    (C) VIEWPORT
    (D) VPLAYER

Level II Objective 5.02

33. You have placed a dimension using the STANDARD style. Click on the area of the Dimension Edit menu illustrated at the right designed to change it to MECHANICAL.

Level II Objective 8.01

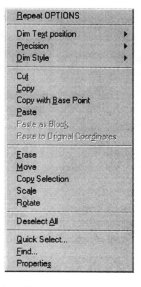

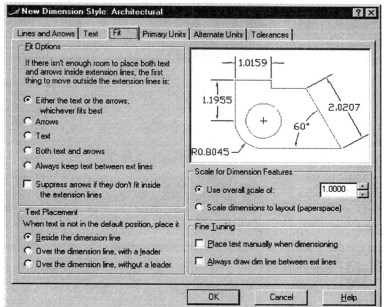

34. You have created floating viewports of various sizes to hold drawing information. Each viewport is assigned a different scale factor. Click on the area of the New Dimension Style dialog box illustrated above designed to control the scale of dimensions based on the viewport scale.

Level II Objective 5.04

35. In the illustration at the right, using the Extend Edge option of the TRIM command will allow how many lines to be trimmed to cutting edge "A"?

    (A) 3 lines
    (B) 4 lines
    (C) 5 lines
    (D) 6 lines

Level II Objective 2.07

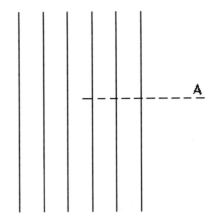

36. What action do you perform to display the special cursor menu in the illustration at the right?

    (A) Right click on a selected viewport.
    (B) Right click in floating model space.
    (C) Right click on one of the paper margins in a layout.
    (D) Right click in any blank part of the screen in layout mode.

Level II Objective 5.05

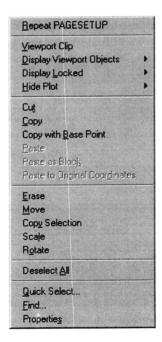

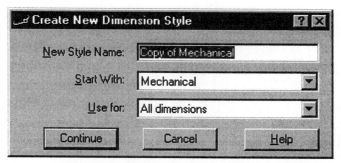

37. You need to place diameter dimensions and have the dimension line forced through the entire diameter of the circle. All diameter dimensions need to share this condition. Rather than creating a separate dimension style, click on the area of the Create New Dimension Style dialog box to begin the process.
Level II Objective 3.02

38. To have AutoLISP routines automatically load when starting AutoCAD, you could collect all routines and place them in a special file. What is the name of this file?
(A) ACAD.LIN
(B) ACAD.LSP
(C) ACAD.PAT
(D) ACAD.PGP
Level II Objective 9.03

39. You attach an external reference called "ASSEMBLY" and examine the current layers. From the following, what would a typical layer name consist of that belongs to the "ASSEMBLY" external reference?
(A) ASSEMBLY/NUT
(B) ASSEMBLY|BOLT
(C) ASSEMBLY#SCREW
(D) ASSEMBLY$SPRING
Level II Objective 7.07

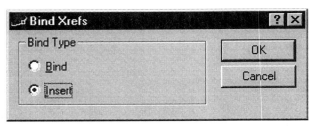

40. An externally referenced drawing called HOUSE.Dwg contains a block called
    DR36. You click on the Bind... button of the External Reference dialog box and
    the Bind Xrefs dialog box illustrated above appears. You click on the Insert
    button. The name of the block becomes
        (A) DR36
        (B) HOUSE-DR36
        (C) HOUSE|DR36
        (D) HOUSE/DR36
Level II Objective 7.08

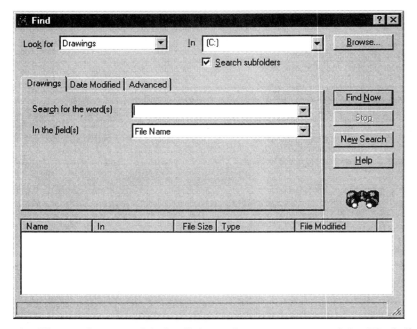

41. From the illustration provided, click on the proper area of the Find dialog box
     designed to tell the search engine to perform a search by layer name.
Level II Objective 4.03

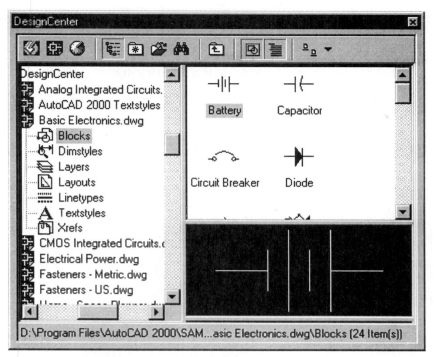

42. Click on the proper area of the DesignCenter dialog box designed to enlarge the image of the battery symbol similar to the illustration provided above.
Level II Objective 4.01

43. You create an electronic plot of a drawing to be shared over the Internet. This drawing will be examined by others using the Autodesk Whip Driver. What will these individuals be prevented from doing to this file?
   (A) Plotting the drawing.
   (B) Modifying the drawing.
   (C) Turning layers on and off.
   (D) Magnifying or demagnifying the drawing.
Level II Objective 5.08

Start a new drawing called Duplex. Begin the construction of the Duplex illustrated below by changing the system of units from decimal to architectural. Keep the remaining default unit settings. No special limits need be set for this drawing. Do not add any dimensions to this drawing. **All wall thicknesses measure 4".** Answer questions 44 and 45 regarding this drawing.

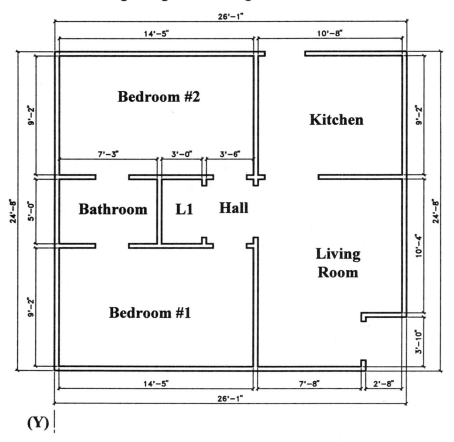

44. Using the illustration of the floor plan above, what is the total combined area of Bedroom #1, Bedroom #2, the Bathroom, the Living Room, and the Laundry Room (L1)?

    (A) 450 sq. ft.
    (B) 458 sq. ft.
    (C) 466 sq. ft.
    (D) 474 sq. ft.

Level II Objective 2.12

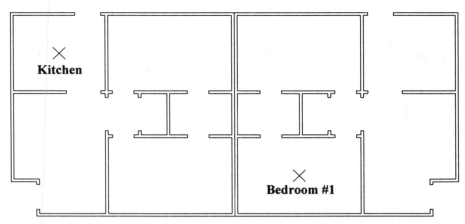

45. From the illustration of the floor plan on the previous page, create a duplex apartment by coping and fliping all objects that make up the single apartment using the centerline "Y" in the illustration on the previous page as the mirror line. Your display should be similar to the illustration above. What is the total distance from the center of the kitchen located in the left duplex to the center of Bedroom #1 located in the right duplex?

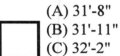

    (A) 31'-8"
    (B) 31'-11"
    (C) 32'-2"
    (D) 32'-5"

Level II Objective 2.02

46. Click on the button in the Refedit Toolbar illustrated above designed to prompt you to select an external reference to edit. This will then launch the Reference Edit dialog box.

Level II Objective 7.04

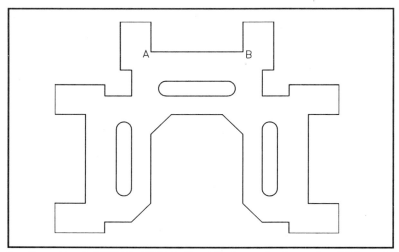

47. Open the drawing METAL STAMP. Reduce the length of Line AB so it has a new total length of 2.12 units. The remainder of the objects should also be reduced proportionally to this line. When completed, what is the new area with the three slots removed?

    (A) 15.07
    (B) 15.11
    (C) 15.15
    (D) 15.19

Level II Objective 2.04

48. You have embedded a Microsoft Word document into an AutoCAD drawing.   In the illustration at the right, click on the area designed to launch an additional dialog box that will allow you to change the scale of the embedded Word document.

Level II Objective 7.03

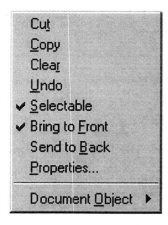

49. You need to redefine an attribute definition. From the illustration, click on the area of the Properties window designed to assign a hidden property to the attribute.

Level II Objective 7.09

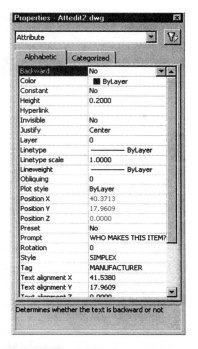

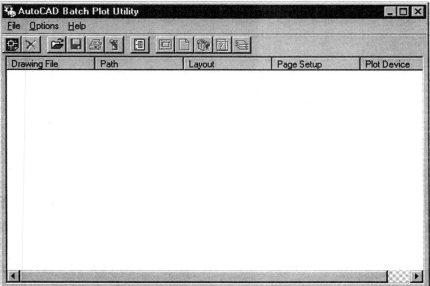

50. Click on the button of the AutoCAD Batch Plot Utility dialog box illustrated above designed to add the names of various drawings to the Batch Plot list.

Level II Objective 5.09

Answers to the AutoCAD 2000 Level II Practice Test

1. A

2. B

3. B

4. C

5. D

6. D

7. Click on the radio button adjacent to "Use color dependent plot styles".

8. C

9. D

10. D

11. A

12. A

13. B

14. B

15. D

16. A

17. B

18. A

19. B

20. Place a check in the box adjacent to "Invisible" in the Mode area of the Attribute Definition dialog box.

21. C

22. C

23. Click on this button:

24. D

25. D

26. C

27. C

28. B

29. D

30. D

31. A

32. D

33. Click on Dim Style and choose a different dimension style.

34. Click on the radio button adjacent to "Scale dimensions to layout (paperspace)".

35. D

36. A

37. Click in the box adjacent to "Use for:" and change All dimensions to Diameter dimensions.

38. B

39. B

40. A

41. Click in the "Look for:" box and change "Drawings" to "Layers".

42. Click on this button:

43. B

44. B

45. A

46. Click on this button:

47. A

48. Click on Properties...

49. Click on the word "Invisible" and change No to Yes.

50. Click on this button:

Notes

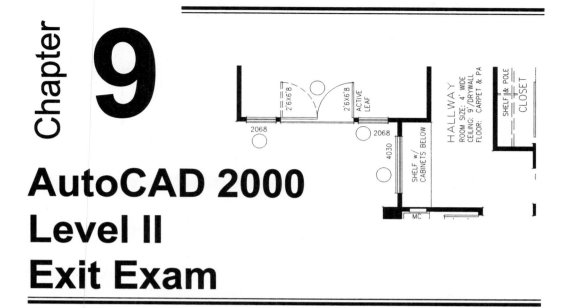

AutoCAD 2000
Level II
Exit Exam

The AutoCAD 2000 Level II Exit Exam has been developed to provide more practice in preparing for the Level II Assessment Exam. The Exit Exam consists of 50 drawing and general knowledge questions. There is no time limit to complete this assessment exam. However, this exam is designed to be completed in 2 hours or less, which would demonstrate use of the software in a productive manner.

Question types include single-answer multiple-choice and hot-spot areas. Numerous questions have been designed around actual images an individual would be confronted with in the production drawing environment.

Two types of drawings are present in this exit exam. Most of the drawings are already created up to a certain point. For these cases, open the drawing file and follow the steps that direct you to perform certain operations before attempting to answer any of the questions that relate to the drawing. All drawings to open are provided on the disk supplied with this manual. Create a folder called \ASSESS-MENT and load all drawing files there. Another type of drawing requires you to construct a new object from the image provided and answer the questions that follow the drawing to test your accuracy.

Work through the Exit Exam at a good pace, paying strict attention to the amount of time spent on each question. Answers for all Level II Exit Exam questions are located at the end of this chapter.

Notes

Provide the best answer for each of the following questions.

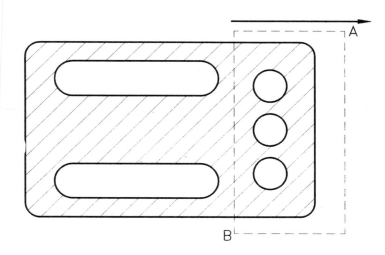

1. The object illustrated above was crosshatched using the BHATCH command. You need to stretch the right side of the object along with the three holes to the right. A crossing box from "A" to "B" is used to capture the items to stretch. What effect does this action have on the crosshatch pattern?

    (A) The objects stretch but the hatch pattern remains unchanged.
    (B) The objects stretch and the hatch pattern updates to this change.
    (C) The hatch pattern stretches but the three holes remain unchanged.
    (D) The objects cannot be stretched once associated with a hatch pattern.

Level II Objective 2.11

2. Start a new drawing. Draw a line segment with the starting point at 2.00,2.00. Construct the next point of the line 23.63 units along the X axis and 4.76 units along the Y axis from the last point. When finished, what is the angle formed by this line segment?

    (A) 9 degrees
    (B) 11 degrees
    (C) 13 degrees
    (D) 15 degrees

Level II Objective 1.01

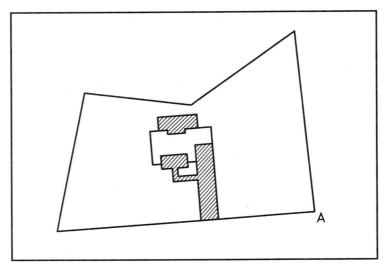

3. Open the existing drawing called PLAT PLAN. Stretch the intersection at vertex "A" a distance of 30' directly to the right. What is the total combined area of the deck, porch, sidewalk, and driveway (all areas identified by hatching)?

    (A) 3352 sq. ft.
    (B) 3355 sq. ft.
    (C) 3358 sq. ft.
    (D) 3361 sq. ft.

Level II Objective 2.05

Details

Name:	HIDDEN	Global scale factor: 1.0000
Description:	Hidden _ _ _ _ _ _ _ _ _ _	Current object scale: 1.0000
☐ Use paper space units for scaling		ISO pen width: 1.0 mm ▼

OK   Cancel   Help

4. Illustrated above is a partial image of the Linetype Manager dialog box. Click in the area designed to affect the scale of linetypes similar to that of using the PSLTSCALE system variable.

Level II Objective 5.03

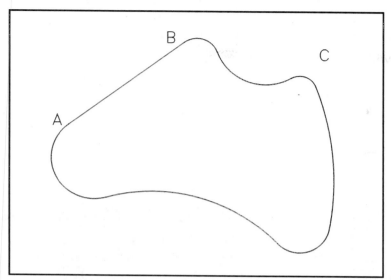

5. Open the existing drawing called SPECIAL CAM. Turn on layers "Object" and "Dim". Using the center of arc "A" as the base point, rotate the profile of the Special Cam so that line "AB" measures 82 degrees. When finished, what is the XY coordinate value of the center of arc "C"?

    (A) 8.133,6.625
    (B) 8.137,6.629
    (C) 8.141,6.629
    (D) 8.141,6.633

Level II Objective 2.03

6. Start a new drawing. Construct a rectangle with the first corner beginning at coordinate 3.77,3.12. Construct the other corner 4.42 units at a 22 degree angle from the last point. When finished, what is the perimeter of the rectangle?

    (A) 11.5072
    (B) 11.5075
    (C) 11.5078
    (D) 11.5081

Level II Objective 1.02

7. You have collected a series of objects using the GROUP command. Whenever a member of the group is selected, all members of the group highlight. What toggle will turn group mode on or off?

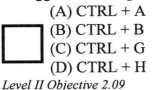
    (A) CTRL + A
    (B) CTRL + B
    (C) CTRL + G
    (D) CTRL + H

Level II Objective 2.09

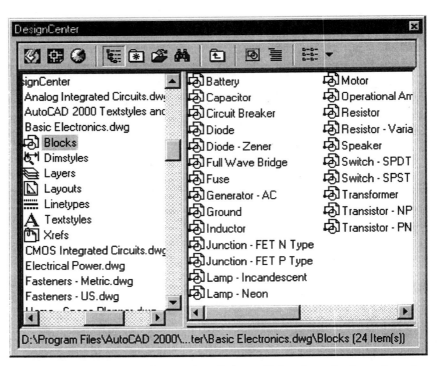

8. Click on the proper area of the DesignCenter dialog box designed to change the current listing of icons to large icons.

Level II Objective 4.01

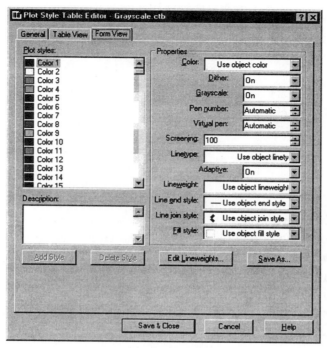

9. Click on the proper tab of the Plot Style Table Editor dialog box designed to add an overall description to the Grayscale.ctb style. This tab also contains information such as the total number of styles and the path where Grayscale.ctb is located.
Level II Objective 5.07

10. You desire to create a non-rectangular viewport. Click on the button in the Viewports toolbar designed to create a viewport in the shape of a multisided polygon.
Level II Objective 5.06

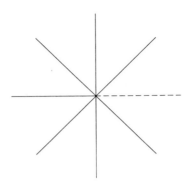

11. Illustrated above is an object being constructed using the ARRAY command
    and selecting the highlighted object. Without using the ARRAY command,
    what alternate method could be used to construct the objects above?
       (A) using the ROTATE command
       (B) using Grip-Rotate mode
       (C) using Grip-Rotate mode with the Multiple option
       (D) using Grip-Rotate mode with the Multiple option and the temporary
           snap mode through the use of the SHIFT key
Level II Objective 2.13

12. You make changes to an AutoCAD menu template file. Before you can use this
    menu, it must first be compiled. What extension does this compiled menu file
    have?
       (A) .mnc
       (B) .mnl
       (C) .mns
       (D) .mnu
Level II Objective 9.02

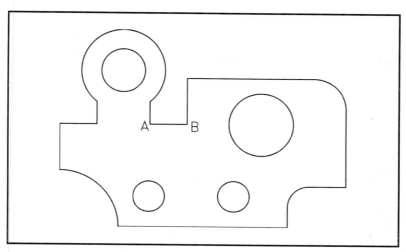

13. Open the drawing LOCATER. Increase the length of Line AB so it has a new total length of 1.75 units. The remainder of the objects should also increase in proportion to this line. When completed, what is the new area with the four holes removed?

    (A) 72.68
    (B) 72.72
    (C) 72.76
    (D) 72.80

Level II Objective 2.04

14. What signifies that a line or an arc has been selected for use by the Object Snap Extension mode?

    (A) A circle displays at the endpoint.
    (B) A triangle displays at the endpoint.
    (C) A square is displayed at the endpoint.
    (D) A small plus sign (+) is displayed at the endpoint.

Level II Objective 6.02

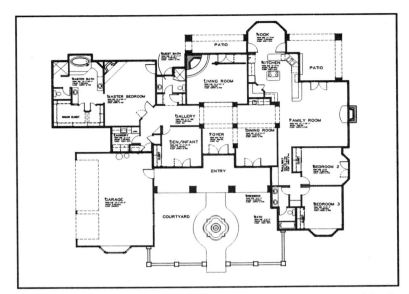

15. Open the existing drawing called HOUSE PLAN. How many blocks identified by the name "RMNBIG" are found in the entire database of this drawing?

    (A) 4
    (B) 8
    (C) 12
    (D) 16

Level II Objective 6.01

16. One feature of the AutoCAD DesignCenter is to search for content. This content can then be dragged and dropped into drawings very efficiently. From the various lists provided, what items would be considered valid content in a search?

    (A) Blocks, Named Views, Xrefs, Layers
    (B) Layers, Linetypes, Textstyles, Xrefs
    (C) Layouts, Layers, Named Views, Drawing Files
    (D) Named Views, Dimension Styles, Drawing Files, Layouts

Level II Objective 4.03

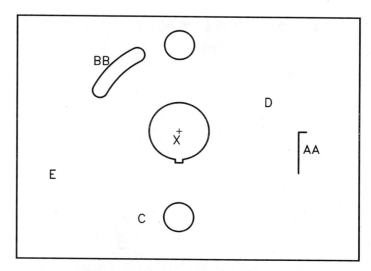

17. Open the existing drawing called RATCHET and perform the following operations: Duplicate the tooth identified as "AA" 18 times in a complete circle. Use the intersection at "X" as the center of the duplication. When finished, what is the total perimeter of the Ratchet?

    (A) 30.78
    (B) 30.82
    (C) 30.86
    (D) 30.90

Level II Objective 2.01

18. You enter the TRIM command. At the "Select objects:" prompt, you press the ENTER key. What happens as a result of this action?

    (A) You are returned to the Command: line.
    (B) You are prompted to select the objects to trim.
    (C) The prompt "Invalid" appears at the Command: line.
    (D) The "Select objects:" prompt displays again at the Command: line.

Level II Objective 2.07

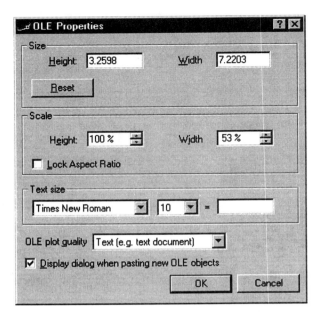

19. You have embedded a Microsoft Word document into an AutoCAD drawing. Unfortunately, the OLE document is not sized proportionally. Click in the area of the OLE Properties dialog box illustrated above designed to make the size of the OLE document proportional.
Level II Objective 7.03

20. You need to document a script file. This comes in the form of adding a comment about the script. AutoCAD will ignore this comment when it executes the script. Which of the following is the correct form of a comment?
    (A) This script is designed to create a number of layers
    (B) :This script is designed to create a number of layers
    (C) ;This script is designed to create a number of layers
    (D) >This script is designed to create a number of layers
Level II Objective 9.04

21. Click on the area of the Dimension Edit menu in the illustration designed to place the dimension value above the dimension line.

Level II Objective 8.01

22. You create the following tool button macro:

| **INSERT;FLANGE;\;;;** |

What does the presence of the back slash "\" signify?

(A) A comment
(B) An ENTER key
(C) Pause for user input
(D) The continuation of the macro

Level II Objective 9.01

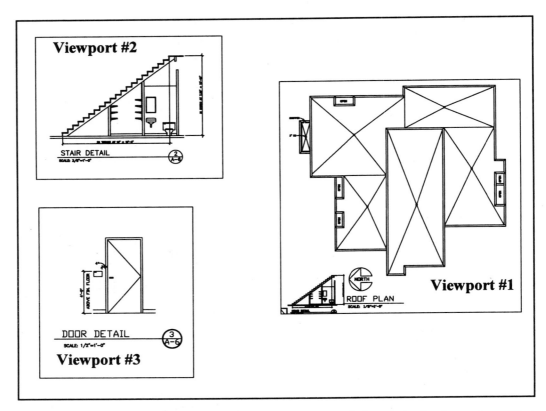

23. Illustrated above are three viewports arranged in a single layout. What command is used to remove the unnecessary images of the stair detail and door detail from Viewport #1?

      (A) MOVE
      (B) PURGE
      (C) SCALE
      (D) VPLAYER

Level II Objective 5.02

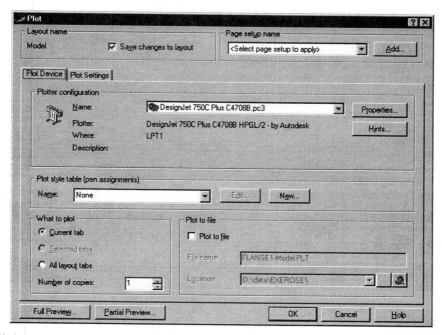

24. Click on the area of the Plot dialog box illustrated above designed to prepare a drawing to be electronically plotted for viewing over the Internet.
Level II Objective 5.08

25. Click on the button in Refedit Toolbar illustrated at the right that will display the AutoCAD alert box warning you that the changes you made to an external reference will be permanently saved to the referenced drawing.
Level II Objective 7.04

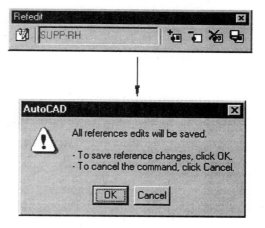

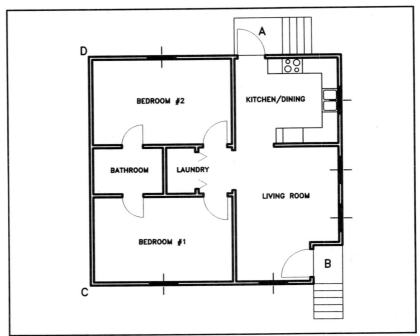

26. Open the existing drawing called DUPLEX. Create a mirror image of the apartment to create the duplex. Use the intersection points of "C" and "D" as mirror points. What is the total distance from the center of concrete pad "A" on the left half of the duplex to the center of concrete pad "B" on the right half of the duplex?
    - (A) 48'-6"
    - (B) 48'-10"
    - (C) 49'-2"
    - (D) 49'-6"

Level II Objective 2.02

27. You have the ability to have custom AutoLISP commands load into your drawing file automatically. What is the file name used to accomplish this task?
    - (A) ACAD.LIN
    - (B) ACAD.LSP
    - (C) ACAD.PAT
    - (D) ACAD.PGP

Level II Objective 9.03

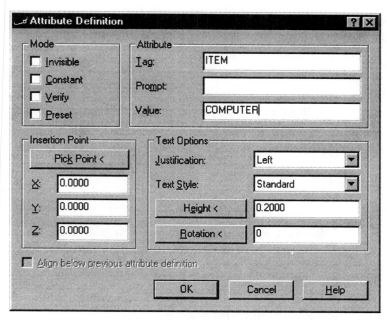

28. You create an attribute definition with a tag of "ITEM" and a value of "COMPUTER." This attribute definition will be assigned to a block. What mode of the Attribute Definition dialog box allows an attribute value to be fixed for all block insertions?

   (A) Constant
   (B) Invisible
   (C) Preset
   (D) Verify

Level II Objective 7.01

29. You are creating a drawing that will be used as an external reference by other designers. The drawing you are creating uses several external references which you do not want to share with others when they xref your drawing. What option of the External References dialog box should you select when you attach external references to your current drawing?

   (A) Attach
   (B) Bind
   (C) Overlay
   (D) Reload

Level II Objective 7.06

30. You have attached the file "PUMP" as an external reference into the current drawing file. Activating the Layer Properties Manager dialog box, you notice the following layer listed: PUMP|OBJECT. From the following list of layer functions, which item has no effect on a layer belonging to an external reference?

(A) Freezing the layer.
(B) Turning the layer off.
(C) Making the layer current.
(D) Changing the color of the layer.

Level II Objective 7.07

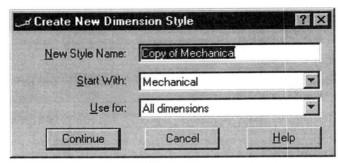

31. You want all radius dimensions to share the same dimension variable settings. In the illustration above, click on the area of the Create New Dimension Style dialog box to accomplish this task.

Level II Objective 3.02

32. You perform a task that suppresses the display and regeneration of the xref definition. This leads to an improvement in the performance of the drawing file without removing the xref from the drawing. What external reference option is used to perform this task?

(A) Bind
(B) Detach
(C) Erase
(D) Unload

Level II Objective 7.02

33. In the Properties window illustrated at the right, the attribute tag "Cost" has been assigned the prompt "How much does this item cost?" Click on the area of this window designed to repeat the attribute prompt when the block is inserted.

Level II Objective 7.09

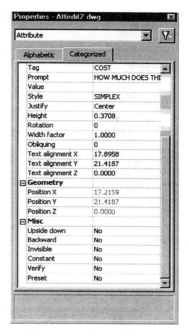

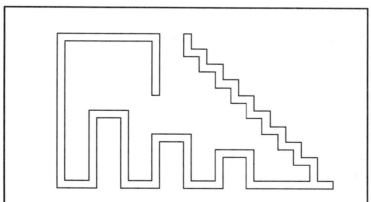

34. Open the drawing called SPRING MOUNT. The drawing should appear similar to the illustration above. Fillet every corner of the extrusion with a radius of 0.10 units. When finished, what is the total area of the Spring Mount?

    (A) 9.1152
    (B) 9.1158
    (C) 9.1164
    (D) 9.1170

Level II Objective 2.08

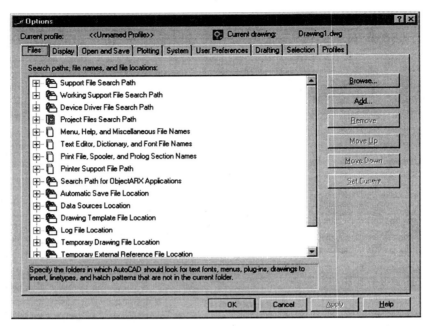

35. You make numerous changes under various tabs of the Options dialog box. These changes can be saved to a user-defined name. Click on the proper tab of the Options dialog box that takes you to the area to perform this task.
Level II Objective 8.03

36. You activate a floating viewport and scale an image to Paper Space units. What should the DIMSCALE system variable be set to for AutoCAD to automatically compute the dimension scale factor based on the scale of the current floating model space viewport?

    (A) 0
    (B) 1
    (C) 2
    (D) 5

Level II Objective 5.04

37. You enter the following relative coordinate: @0,-27. What does the presence of the "@" symbol mean?

(A) 0,0
(B) origin
(C) last point
(D) next point

Level II Objective 1.04

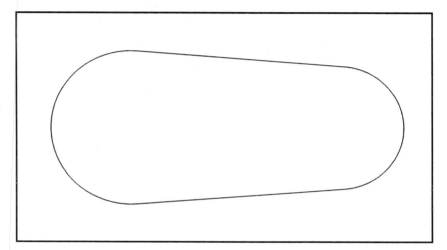

38. Open the existing drawing called CHAIN LINKS. A block called LINKS already exists in the drawing. Insert this block at 0.50 increments along the perimeter of the polyline shape in the illustration above. Be sure to align the block with the polyline shape. When finished, how many blocks called LINKS are actually inserted?

(A) 23
(B) 32
(C) 44
(D) 50

Level II Objective 1.03

Start a new drawing called Gusset. Begin the construction of the Gusset illustrated below by keeping the default units set to decimal but change the number of decimal places past the zero from 4 to 2. No special limits need be set for this drawing. Do not add any dimensions to this drawing. Answer question #39 using this drawing.

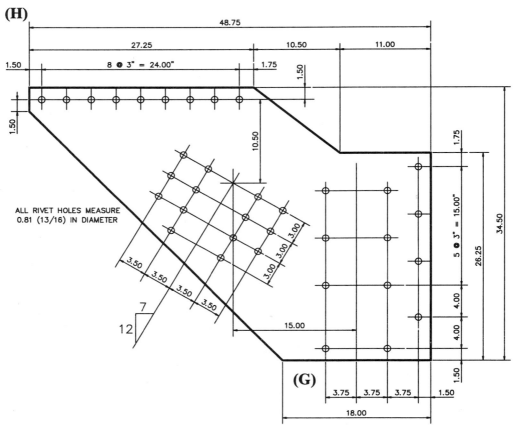

39. Stretch the Gusset plate directly to the left using a crossing box from "G" to "H" and at a distance of 3.00 units. What is the new area of the Gusset plate with all 35 rivet holes removed?

    (A) 1121.98
    (B) 1126.72
    (C) 1131.11
    (D) 1136.48

Level II Objective 2.12

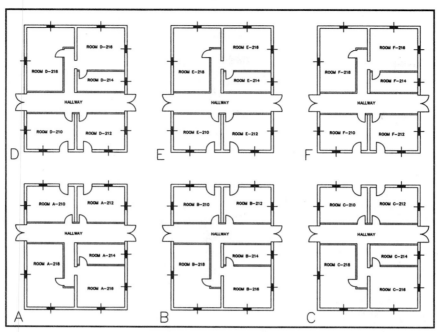

40. Open the existing drawing called BUILDINGS. How many layers begin with the letter L, are assigned the color yellow, and are turned off in the database of this drawing?

    (A) 1
    (B) 3
    (C) 11
    (D) 16

Level II Objective 6.03

41. Numerous attributes are present in the current drawing file. You now have to extract the attributes into either the CDF, SDF, or DXF formats. What command is used to perform this task?

    (A) ATTDEF
    (B) ATTDISP
    (C) ATTEDIT
    (D) ATTEXT

Level II Objective 7.05

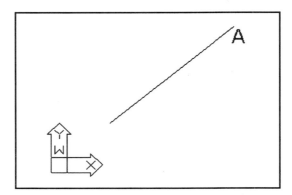

42. Open the drawing file called LINE. Extend the end of the line at "A" so that the total length of the line measures 7.50 units. When finished, what are the Delta X,Y coordinates of the new end of the line segment?

    (A) 5.8469,4.6927
    (B) 5.8479,4.6937
    (C) 5.8489,4.6947
    (D) 5.8499,4.6957

Level II Objective 2.06

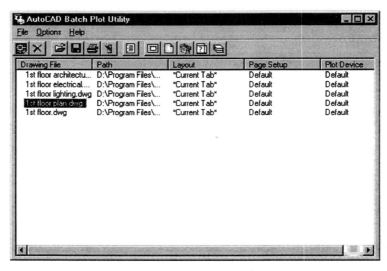

43. Various drawings are listed in the AutoCAD Batch Plot Utility dialog box above. You select one of the drawings. Click on the button in this dialog box designed to view the layers that apply to this selected drawing. This will allow you to specify if you want to plot certain layers or not.

Level II Objective 5.09

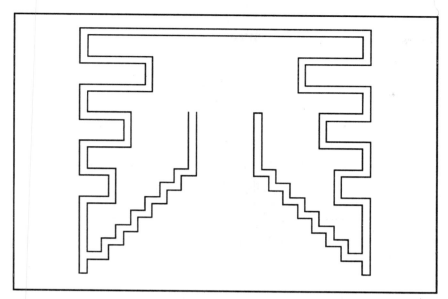

44. Open the existing drawing called MIRROR PATTERN. Convert all lines into one continuous polyline object. When finished, what is the total length of this polyline?

    (A) 171.8113
    (B) 171.8120
    (C) 171.8127
    (D) 171.8134

Level II Objective 2.10

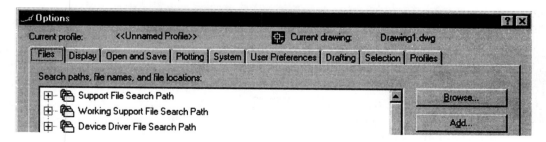

45. Click on the proper tab of the Options dialog box designed to take you to the area to add or configure a new output device.

Level II Objective 8.02

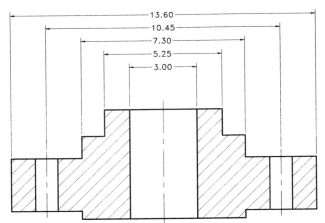

46. You use the QDIM command to dimension the object illustrated above. Which Quick Dimension mode is depicted in this illustration?

    (A) Baseline
    (B) Continuous
    (C) Ordinate
    (D) Staggered

Level II Objective 3.01

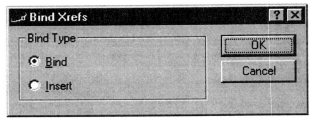

47. You have externally referenced the file "SCHEMATIC" into your current drawing file. You now want to convert the external reference into a block through the Bind option of the XREF command. However the Bind Xrefs dialog box illustrated above appears. When you click on the Bind radio button, the block "RESISTOR" is renamed to

    (A) RESISTOR
    (B) SCHEMATIC|RESISTOR
    (C) SCHEMATIC0RESISTOR
    (D) SCHEMATIC%_%RESISTOR

Level II Objective 7.08

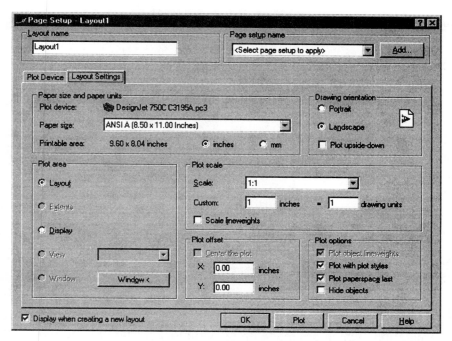

48. Click on the proper radio button of the Page Setup dialog box illustrated above designed to rotate the plot by 90 degrees.

Level II Objective 5.01

49. You activate the Viewports toolbar in the illustration above. Why is the scale of the viewport grayed out?

 (A) The viewport has been locked.

 (B) The viewport must first be active.

 (C) An invalid scale has been entered into this area.

 (D) The scale applies to the drawing in Model Space.

Level II Objective 5.05

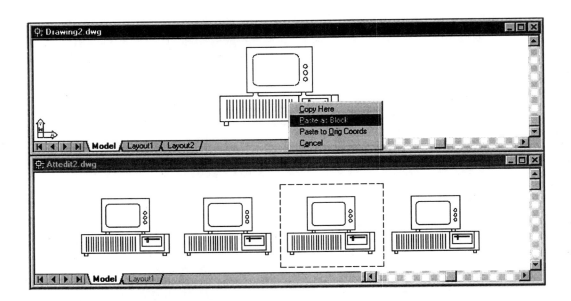

50. Study the illustration above. The bottom viewport consists of various computer
    workstations which are made up of two blocks; MONITOR and CPU. You
    drag and drop one set of these blocks outlined by the dashed rectangle into the
    upper viewport by right clicking. When you perform this operation, a special
    cursor menu appears. If you use the "Paste as Block" option, what is the new
    name for the pasted block?

    (A) An arbitrary name is assigned.
    (B) CPU
    (C) MONITOR
    (D) X1

Level II Objective 4.02

Answers to the AutoCAD 2000 Level II Exit Exam

1. B

2. B

3. A

4. Place a check in the box adjacent to "Use paper space units for scaling".

5. A

6. C

7. A

8. Click on this button:

9. Click on the General tab of the Plot Style Table Editor dialog box.

10. Click on this button:

11. D

12. A

13. B

14. D

15. C

16. B

17. A

18. B

19. Place a check in the box adjacent to "Lock Aspect Ratio".

20. C

21. Click on "Dim text position".

22. C

23. D

24. In the Plotter configuration area of the Plot dialog box, click in the box adjacent to "Name:" and change the plotter to "DWF ePlot.pc3".

25. Click on this button:

26. B

27. B

28. A

29. C

30. C

31. Click in the box adjacent to "Use for:" and change "All dimensions" to "Radius dimensions".

32. D

33. In the Misc area of the Properties window, click on "Verify" and change No to Yes.

34. C

35. Click on the Profiles tab of the Options dialog box.

36. A

37. C

38. C

39. D

40. A

41. D

42. C

43. Click on this button:

44. A

45. Click on the Plotting tab of the Options dialog box.

46. D

47. C

48. In the "Drawing orientation" area of the Plot Setup dialog box, click on the radio button adjacent to "Portrait".

49. A

50. A

10

AutoCAD LT 2000
Level I and Level II Exams
Categories and Objectives

The AutoCAD LT 2000 Assessment Exams consist of drawings and general knowledge questions that cover various AutoCAD topics. Inquiry commands are used to analyze each drawing question. This takes the form of using such commands as AREA, DIST, ID, and LIST for performing various calculations on each drawing. Knowledge of using the Properties dialog box and the ability to create selection sets using Quick Select would also be helpful.

General knowledge questions may take the form of the following question types:

> Single answer multiple choice
> Hot spot

This chapter outlines the categories that make up the AutoCAD LT 2000 Level I and Level II Exams complete with the number of questions and a topic percentage that relates to the entire exam. Each category is further outlined with a detailed listing of the objectives an individual must master to be successful with the AutoCAD LT 2000 Exams.

Notes

AutoCAD LT 2000
Level I Exam Categories

The AutoCAD LT 2000 Level I Exam consists of single answer multiple choice questions and hot spot questions. Use the chart below for a breakdown on the question categories, the number of question per category, and the weight they carry in the AutoCAD LT 2000 Level I Assessment Exam.

AutoCAD LT 2000 Level I Exam Categories	Number of Questions	Percentage of Overall Score
1 Creating Drawing Template Files	2	4%
2 Display Commands	2	4%
3 Drawing Objects	12	24%
4 Extracting Drawing Information	1	2%
5 Editing	17	34%
6 Annotating Drawings	6	12%
7 Dimensioning a Drawing	3	6%
8 Managing Content	4	8%
9 Plotting	3	6%
Total	**50**	**100%**

AutoCAD LT 2000
Level I Exam Objectives

Category 1
Creating Drawing Template Files
Obj. 1.01   Create and delete layers
Obj. 1.02   Set layer properties

Category 2
Display Commands
Obj. 2.01   Use all options of the ZOOM command
Obj. 2.02   Save and restore a named view

Category 3
Drawings Objects
Obj. 3.01   Draw objects using Absolute Coordinates
Obj. 3.02   Draw objects using Relative Coordinates
Obj. 3.03   Draw objects using Polar Coordinates
Obj. 3.04   Use Direct Distance entry
Obj. 3.05   Use Polar Tracking
Obj. 3.06   Change Polar Tracking settings
Obj. 3.07   Draw a circle using all options
Obj. 3.08   Draw an arc using all options
Obj. 3.09   Set Point Size and Appearance
Obj. 3.10   Use the MEASURE and DIVIDE commands
Obj. 3.11   Use and identify Object Snaps
Obj. 3.12   **Draw an ellipse***

*Note: The objective in bold is in the AutoCAD LT 2000 exam only.

Category 4
Extracting Drawing Information
Obj. 4.01   Add and subtract areas using the AREA command

Category 5
Editing
Obj. 5.01   Select objects by Fence
Obj. 5.02   Use Previous and Last options
Obj. 5.03   Use Object Cycling
Obj. 5.04   Copy and move objects
Obj. 5.05   Use the OFFSET command
Obj. 5.06   Use the ARRAY command
Obj. 5.07   Use the MIRROR command
Obj. 5.08   Use the ROTATE command
Obj. 5.09   Use the SCALE command
Obj. 5.10   Use the STRETCH command
Obj. 5.11   Use the EXTEND command
Obj. 5.12   Use the TRIM command
Obj. 5.13   Use the FILLET and CHAMFER commands
Obj. 5.14   Use the BREAK command
Obj. 5.15   Use Grips to edit objects
Obj. 5.16   Use PEDIT to convert and join objects into a pline
Obj. 5.17   Use the Properties window

Category 6
Annotating Drawings
Obj. 6.01   Use the STYLE command, set and edit a text style
Obj. 6.02   Create single line and multiline text
Obj. 6.03   Format text using MTEXT
Obj. 6.04   Use the spell checker
Obj. 6.05   Use the BHATCH command
Obj. 6.06   Edit a hatch pattern

*Note: The objective in bold is in the AutoCAD LT 2000 exam only.

Category 7
Dimensioning a Drawing
Obj. 7.01   Use QDIM and DIMLINEAR
Obj. 7.02   Use the QLEADER command and its options
Obj. 7.03   Use the Dimension Style Manager

Category 8
Managing Content
Obj. 8.01   Create, insert and redefine a block
Obj. 8.02   Use MDE to copy objects and properties between drawings
Obj. 8.03   View and copy content with DesignCenter
Obj. 8.04   Use the PURGE command to reduce drawing size

Category 9
Plotting
Obj. 9.01   Setup a plot
Obj. 9.02   Use plot scale in model space and layouts
Obj. 9.03   Rename a layout

*Note: The objective in bold is in the AutoCAD LT 2000 exam only.

AutoCAD LT 2000
Level II Exam Categories

The AutoCAD LT 2000 Level II Exam consists of single answer multiple choice questions and hot spot questions. Use the chart below for a breakdown on the question categories, the number of question per category, and the weight they carry in the AutoCAD LT 2000 Level II Assessment Exam.

AutoCAD LT 2000 Level II Exam Categories	Number of Questions	Percentage of Overall Score
1 Drawing Objects	4	8%
2 Editing	13	26%
3 Dimensioning	2	4%
4 Managing Content	3	6%
5 Plotting	9	18%
6 Object, Layer, and Point Control	3	6%
7 Attributes, External References and Images	9	18%
8 Advanced Preference Settings and System Variables	3	6%
9 Customization	4	8%
Total	**50**	**100%**

AutoCAD LT 2000
Level II Exam Objectives

Category 1
Drawing Objects
Obj. 1.01   Draw objects using Relative Coordinates
Obj. 1.02   Draw objects using Polar Coordinates
Obj. 1.03   Use the MEASURE and DIVIDE commands
Obj. 1.04   Know the meaning of the @ symbol

Category 2
Editing
Obj. 2.01   Use the ARRAY command
Obj. 2.02   Use the MIRROR command
Obj. 2.03   Use the ROTATE command
Obj. 2.04   Use the SCALE command
Obj. 2.05   Use the STRETCH command
Obj. 2.06   Use the LENGTHEN command
Obj. 2.07   Use the TRIM command
Obj. 2.08   Use the FILLET and CHAMFER commands
Obj. 2.09   Group objects
Obj. 2.10   Use PEDIT to convert and join objects into a pline
Obj. 2.11   Edit a Hatch Pattern
Obj. 2.12   Add and subtract areas using the AREA command
Obj. 2.13   Apply linear or polar snaps using grips with the SHIFT key

Category 3
Dimensioning
Obj. 3.01   **Place rotated dimensions using the DIMLINEAR command.***
Obj. 3.02   Use the Dimension Style Manager

*Note: The objective in bold is in the AutoCAD LT 2000 exam only.

Category 4
Managing Content
Obj. 4.01   View and copy content with DesignCenter
Obj. 4.02   Transfer data using the Multiple Drawing Environment
Obj. 4.03   Perform searches for content using AutoCAD DesignCenter

Category 5
Plotting
Obj. 5.01   Design a layout
Obj. 5.02   Control layer visibility by viewport
Obj. 5.03   Use PSLTSCALE to control linetype scaling in different floating viewports
Obj. 5.04   Scale dimensions in different floating viewports using a DIMSCALE of 0
Obj. 5.05   Know the purpose and function of locking viewports
Obj. 5.06   **Know the purpose of a .PC3 file.***
Obj. 5.07   Create and use Plot styles
Obj. 5.08   Know how to create and view .DWF files
Obj. 5.09   **Apply different scale factors to viewports for plotting.***

Category 6
Object, Layer and Point Control
Obj. 6.01   **Create a selection set using the QSELECT command.***
Obj. 6.02   **Know the limitations of layer naming***
Obj. 6.03   Apply filters to layers

*Note: The objective in bold is in the AutoCAD LT 2000 exam only.

Category 7
Attibutes, External References and Images
Obj. 7.01   Understand the purpose and use of elements of attribute defini-
             tions
Obj. 7.02   Use the External Reference Manager
Obj. 7.03   Use the OLE scaling capability
Obj. 7.04   **Edit attributes individually using DDATTE.***
Obj. 7.05   Extract attribute information using DDATTEXT
Obj. 7.06   Distinguish between attached and overlaid XREFs
Obj. 7.07   Identify the syntax of layer names of XREFs that are bound or
             attached
Obj. 7.08   Distinguish between bind and insert options in the Bind Xrefs
             dialog box
Obj. 7.09   Redefine attributes using the Properties window

Category 8
Advanced Preference Settings and System Variables
Obj. 8.01   Use the Dimension Edit menu
Obj. 8.02   Know the functions of the Plotting tab of the Options dialog box
Obj. 8.03   **Know the functions of the User Preferences tab of the Options**
             dialog box.*

Category 9
Customization
Obj. 9.01   Create a new tool button macro
Obj. 9.02   Know the purpose of the menu files .mnu, .mns, .mnc
Obj. 9.03   **Identify the proper syntax for menu files.***
Obj. 9.04   Identify the syntax for script files

*Note: The objective in bold is in the AutoCAD LT 2000 exam only.

The following sample question applies only to the AutoCAD LT 2000 Level I Assessment exam:

1. Enter AutoCAD and create a new drawing. Set the units to 4 decimal place accuracy. Construct an ellipse object. The major axis of the ellipse is 9.9860 units. The minor axis is 6.1270 units. When finished, what is the measurement of the perimeter of the ellipse?
    (A) 25.6739
    (B) 25.6745
    (C) 25.6751
    (D) 25.6757

AutoCAD LT 2000 Level I - Objective 3.12

The following sample questions apply only to the AutoCAD LT 2000 Level II Assessment exam:

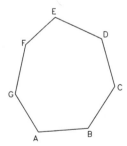

1. Open the drawing PERIMETER. Dimension the distance from the intersection at "A" to the intersection at "D". This linear dimension needs to be placed at an angle of 47 degrees. When finished, what is the length of this distance?
    (A) 4.8266
    (B) 4.8271
    (C) 4.8276
    (D) 4.8281

AutoCAD LT 2000 Level II - Objective 3.01

2. You create a configured plotter file using the Add-a-Plotter wizard. What extension does this file have?

     (A) PCP
     (B) PC2
     (C) PC3
     (D) PLT

AutoCAD LT 2000 Level II - Objective 5.06

3. What option of the ZOOM command allows you to scale an image inside of a floating viewport to paper space units?

     (A) Dynamic
     (B) Extents
     (C) X
     (D) XP

AutoCAD LT 2000 Level II - Objective 5.09

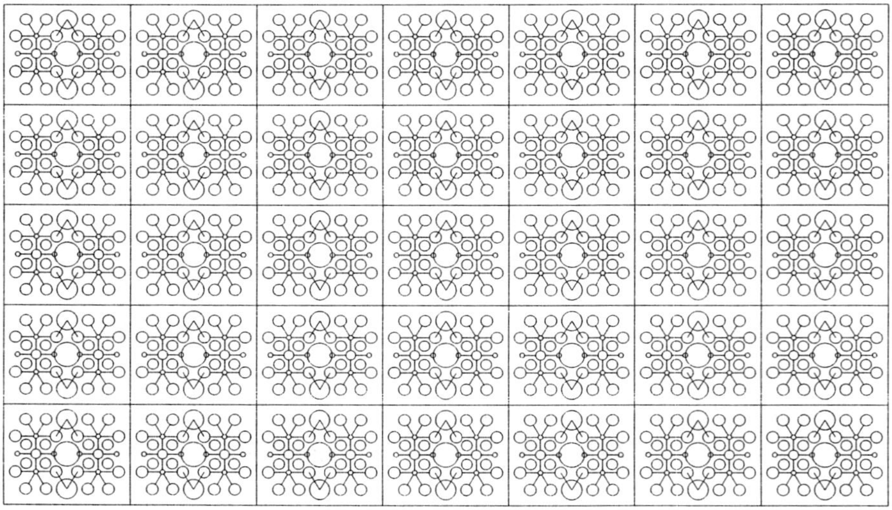

4. Open the drawing COLOR PATTERN1 illustrated above. What is the total number of circles assigned color 30 and that have a radius of less than 0.375?

     (A) 769
     (B) 910
     (C) 1032
     (D) 1153

AutoCAD LT 2000 Level II - Objective 6.01

5. From the following list, which layer name would be considered invalid?
      (A) Mechanical Part
      (B) Mechanical/Part
      (C) Mechanical-Part
      (D) Mechanical_Part

AutoCAD LT 2000 Level II - Objective 6.02

6. You have placed a series of attributes in your drawing. However the values for a few of the attributes need to be changed. What command will activate the Edit Attributes dialog box which will allow you to edit attribute values easily?
      (A) CHANGE
      (B) DDATTE
      (C) DDEDIT
      (D) MODIFY

AutoCAD LT 2000 Level II - Objective 7.04

7. You click on the right mouse button. Instead of simulating the ENTER key, a menu pops up on the screen. What tab of the Options dialog box allows you to control this action of the right mouse button?
      (A) Display
      (B) Open and Save
      (C) System
      (D) User Preferences

AutoCAD LT 2000 Level II - Objective 8.03

8. You want to create a new menu section called POP12. How is this header name identified in a menu file?
      (A) **POP12
      (B) ***POP12
      (C) $POP12
      (D) +POP12

AutoCAD LT 2000 Level II - Objective 9.03

Answers to the AutoCAD LT 2000 Level I Sample Question

1. B

Answers to the AutoCAD LT 2000 Level II Sample Questions

1. C

2. C

3. D

4. B

5. B

6. B

7. D

8. B

Appendix

A

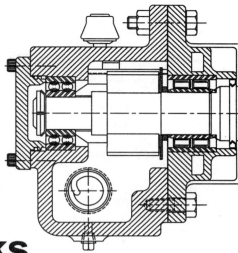

Exam Taking
Tips and Tricks

Use the following hints and suggestions to successfully complete the AutoCAD Assessment exams:

1. Have a good attitude entering the exam. Take the exam with the intent that you will do well.

2. You must be completely comfortable with basic operations performed in Windows 95, Windows 98, or Windows NT 4.0. Know how to switch between different Windows applications. Know how to minimize and maximize a Windows screen. Know also how to size individual screens.

3. You must be completely comfortable with the use of the following Inquiry commands used to assist in answering drawing questions: AREA, DIST, ID, LIST, DDMODIFY

4. Don't be complacent when taking the exam. Attack the exam moving through the drawing and general knowledge questions at a good pace.

5. Do not get bogged down with questions you do not know or are unsure about. Keep track of the questions you are unsure about. Move on with the exam and come back to those questions you are unsure about later.

6. Do not leave any questions unanswered. If you are unsure about a question, narrow down the number of choices to have a better chance of getting the question correct.

7. Since the objectives of each exam are designed from course objectives, attend AutoCAD classes at Authorized AutoCAD Training Centers (ATC). Most training centers offer the following classes or seminars:
    - Level I - Basic Drawing Techniques - the material covered in this course has the same objectives as the Level I Assessment Exam.
    - Level II - Advanced Drawing Techniques - The material covered in this course has the same objectives as the Level II Assessment Exam.

8. Take an exam preparation class at a local Authorized AutoCAD Training Center (ATC.) Various courses are offered Online over the Internet. Contact your local ATC for more information on classes to better prepare for the certification exams.

9. Use this AutoCAD Assessment Exam Prep Manual wisely. Realistically time yourself when taking the Pretest, Practice Test, and Exit Exam in each level. The preparation manual should serve to identify areas in which you may be weak.

10. Use your time wisely during the exam. If time remains, use it to check marked questions, incomplete questions, or questions you are unsure about. It is highly recommended that you use all available time to guarantee your success in the exams.

11. Before working on any drawing, read the question carefully. Draw only what is required to answer the question. For example, layers may be created and used for clarity but may not be required. Also, drawings usually are not required to be dimensioned.

12. Take advantage of using the AutoCAD Help Topics dialog box which is activated with the HELP command to assist in answering General Knowledge questions.

13. Use the power of AutoCAD to test out possible solutions to General Knowledge questions.

B

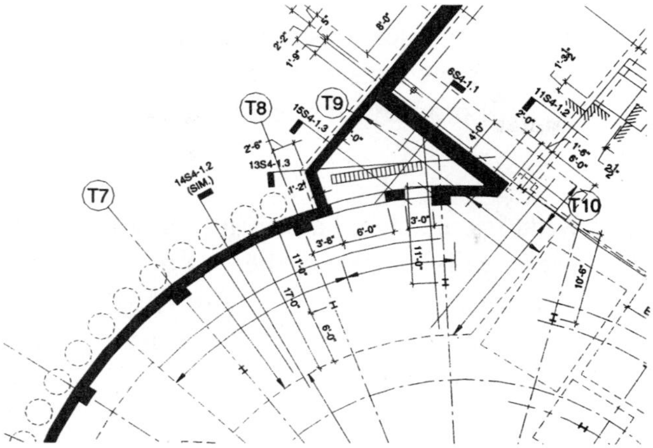

AutoCAD
Assessment Exams FAQ

Q: What are the AutoCAD Assessment Exams?

A: The exams are an Assessment program for AutoCAD professionals. The exams test a user's knowledge and skill with various Autodesk products.

Q: How many AutoCAD assessment exams are available?

A: Assessment exams are available on the following Autodesk products:

> AutoCAD 2000
> AutoCAD LT 2000
> AutoCAD Release 14

Q: How many exams cover each product?

A: Two exams are available for each Autodesk product namely, Level I and Level II. Both exams test a user's knowledge and skill with regards to the specific Autodesk product.

Q: How many questions are asked in each exam?

A: Each exam consists of 50 questions on general knowledge topics and drawing skill productivity exercises.

Q: What is considered a passing score for the exams?

A: Technically speaking, there is no passing score for the Assessment exams. However, a minimum score of 85% in either the Level I or Level II Assessment Exams means you have performed well in most general knowledge and drawing objectives.

Q: Can I take AutoCAD exams on paper?

A: Paper-based exams are not available.

Q: Where are the exams delivered?

A: The exams are delivered on any computer that is Internet connected.   This could be at an Authorized AutoCAD Training Center, at a place of employment, in a college classroom, or in the privacy of your home.

Q: If I do poorly in an Assessment exam, can I take it over again?

A: Yes.   You may take an exam as often as you like.   You will need to re-register (and pay) each time you wish to take an exam.

Q: How much does it cost to take the Assessment exams?

A: Each exam costs US $59.95. Payment is by VISA, MasterCard or American Express only. Purchase orders are not accepted.

Q: How does one register for an exam?

A: You can register for the exam online by visiting www.autodesk.com/exams You will need to be prepared with a credit card number, test taker's contact information, preferred test dates and locations. Telephone registration is not available.

Q: How are exams delivered?

A: The exam is delivered online via a modem and computer.

Q: What types of questions are asked in the Assessment exams?

A: Typical exam questions include multiple choice and "hot spot" questions as well as actual drawing problems.

Q: Are questions designed to be tricky?

A: Questions you will be confronted with are designed to test your Level I and/ or Level II skill level. All efforts have been made to design questions that are non-trivial in nature.

Q: How many general knowledge questions are on the exam?

A: Of the 50 questions total, approximately half are considered general knowledge questions. The remainder consists of drawing productivity questions.

Q: In what format are the general knowledge questions presented?

A: Question types include multiple choice and hot spot. There are no true/false or fill in the blank questions.

Q: How many drawing questions are on the exam?

A: Approximately 25.

Q: In what format are the drawing exercises presented?

A: You will be asked to either open a drawing file, which is provided and make specific changes or draw something from scratch. You will then be asked a multiple-choice question based on the drawing changes. Using an AutoCAD Inquiry command on the changes you made derives the correct answer.

Q: Are drawings plotted out? How are they scored in the exams?

A: Drawings are not plotted out. Multiple choice questions are designed around each drawing problem. Inquiry commands such as AREA, DIST, ID and LIST are used to answer each question. You must be extremely comfortable with each of these commands in order to perform well on the drawing exercise problems. Knowledge of the Properties Window in AutoCAD 2000 and AutoCAD LT 2000 would also be helpful.

Q: What is a Hot Spot question?

A: This type of question allows you to interact with a general AutoCAD question. For example, an image of a typical AutoCAD 2000 dialog box may appear along with a question requiring you to actually click on the correct area of the dialog box. If the correct area is clicked on, a hot spot or hidden area sends a command back to the test generator that the correct answer has been selected. In this way, graphical AutoCAD dialog boxes can be tested as a general question.

Q: Are application-specific exams available (for example, Architecture, Mechanical)?

A: No. At this time, only general AutoCAD exams are available. The drawing files are generic in nature and do not test knowledge of architectural or mechanical design. Assessment exams on other Autodesk products are being developed, however.

Q: What content are the exams based on?

A: The content of each exam is based on Autodesk Official Training Courseware for AutoCAD Release 14, AutoCAD 2000 and AutoCAD 2000 LT. Content is also derived from the Autodesk Training Center course standards for AutoCAD Level I and Level II.

Q: What technical content is covered in the AutoCAD 2000 Level I exam?

A: You will be asked various questions in the Level I exam based on the following categories:

    Creating Drawing and Template Files
    Display Commands
    Drawing Objects
    Extracting Drawing Information
    Editing; Annotating Drawings
    Dimensioning a Drawing
    Managing Content
    Plotting.

Information is also available through the following Web site:

www.autodesk.com/exams

Q: What technical content is covered in the AutoCAD 2000 Level II exam?

A: You will be asked various questions in the Level II exam based on the following categories:

    Drawing Objects
    Editing
    Dimensioning
    Managing Content
    Plotting
    Object, Layer and Point Control
    Attributes
    External References and Images
    Advanced Preference Settings and System Variables
    Customization

Information is also available through the following Web site:

www.autodesk.com/exams

Q: What technical content is covered in the AutoCAD 2000 LT Level I exam?
A: You will be asked various questions in the Level I exam based on the following categories:

      Creating Drawing and Template Files
      Display Commands
      Drawing Objects
      Extracting Drawing Information
      Editing; Annotating Drawings
      Dimensioning a Drawing
      Managing Content
      Plotting

Information is also available through the following Web site:

www.autodesk.com/exams

Q: What technical content is covered in the AutoCAD LT 2000 Level II exam?
A: You will be asked various questions in the Level II exam based on the following categories:

      Drawing Objects
      Editing
      Dimensioning
      Managing Content
      Plotting
      Object, Layer and Point Control
      Attributes
      External References and Images
      Advanced Preference Settings and System Variables
      Customization

Information is also available through the following Web site:

www.autodesk.com/exams

Q: The categories for the AutoCAD 2000 and AutoCAD LT 2000 Assessment exams appear to be identical. Are there any differences in exam content between these two products?

A: While AutoCAD 2000 and AutoCAD LT 2000 Assessment exams are very close in content, there are minor differences in a few objectives between the products. For example, one objective in the AutoCAD 2000 Level I Assessment exam could not be performed in AutoCAD LT 2000. As a result, a new objective was developed to support the AutoCAD LT 2000 Level I exam. In the same manner, eight objectives in the AutoCAD 2000 Level II Assessment exam could not be performed in AutoCAD LT 2000. Eight new objectives were developed to support the AutoCAD LT 2000 Level II exam. Except for these differences, both the AutoCAD 2000 and AutoCAD LT 2000 Assessment exams share a majority of the same objectives.

Q: What technical content is covered in the AutoCAD Release 14 Level I exam?

A: You will be asked various questions in the Level I exam based on the following categories:

AutoCAD Terminology
The Drawing Editor and User Interface
Coordinate Systems
2D Drawing Commands
Drawing Setup & Configuring AutoCAD
Creating & Opening Drawings
Modify Commands
Selection Sets, Object Snaps & Entity Grips
Display Commands
Inquiry Commands
Plotting & Printing Techniques
Blocks
Layer & Object Properties
Dimensioning
Crosshatching
Drawing Annotation
Utility Commands

Information is also available through the following Web site:

www.autodesk.com/exams

Q: What technical content is covered in the AutoCAD Release 14 Level II exam?

A: You will be asked various questions in the Level II exam based on the following categories:

      Level I Review
      Advanced Selection Sets
      Object and Layer Filters
      Multilines; Blocks and Attributes
      Advanced Preference Settings and System Variables
      External References
      Object Linking and Embedding (OLE)
      Raster Image Support
      Paper Space and Model Space
      Advanced Plotting
      Data Management
      Introduction to Customization
      Introduction to ActiveX

Information is also available through the following Web site:

www.autodesk.com/exams

Q: What operating systems are used to take the AutoCAD exams?

A: The AutoCAD 2000 Assessment exams are only offered on the Windows operating system platform. This includes Windows 95, Windows 98, and Windows NT.

Q: Can I use custom menu files and calculators during the exam?

A: Since the exam is a measurement of your present skill level, you are allowed whatever means possible to answer each question. Concurrent use questions in the form of drawing problems are designed to be completed in the most efficient and productive means possible.

Q: How soon do I receive the results of the exam?

A: Your exam results will be automatically posted to your testing site computer immediately upon completing the exam. A score report and a diagnostic evaluation of your performance on the exam is available. Your exam results are kept confidential and can only be viewed by you.

Q: Can I skip the Level I exam and take just the Level II exam to complete Assessment?

A: Yes, you can jump to the Level II exam even though you have not taken the Level I exam. The Level II exam includes various categories that require Level I knowledge to complete successfully.

Q: Why would I want to take the Level I exam?

A: Many people are not prepared to take a Level II exam. The Level I exam provides you with an opportunity to measure your fundamental knowledge and skill with AutoCAD. Being able to provide an employer with evidence of your mastery of fundamental knowledge could prove beneficial for some positions. In addition, Level I is excellent practice and preparation for the Level II.

Q: How do I know if I am ready to take the Level I exam?

A: The Level I exam is recommended for candidates who have technical competency equivalent to 32-hours of intensive instruction on AutoCAD 2000 Level I course content and approximately 300 hours of production time using AutoCAD software.

Q: How do I know if I am ready to take the Level II exam?

A: The Level II exam is recommended for candidates who have technical competency equivalent to 32-hours of intensive instruction on AutoCAD 2000 Level II course content with on-the-job experience or 600 hours of production time using AutoCAD.

Q: Can I take both exams at the same time?
A: Yes, you are welcome to take both on the same day.

Q: How long does an exam take?
A: All Assessment exams do not have a time limit of completion. In a productive sense, all exams should be completed in 2 hours or less.

Q: Should I spend a long time on each of the questions?
A: A good exam taking technique is to move through the exam at a good pace answering only those questions of which you are really sure. You can return later to answer the questions that remain.

Q: Is the test taker required to use AutoCAD during the course of the test?
A: Yes. Approximately 50% of the questions in each exam will test the exam candidate's hands-on experience with the software; therefore, AutoCAD will be used during the actual exam. Test takers will be required to switch to AutoCAD to view or edit existing drawings or create new drawings and then answer questions based on the drawings.

Q: Do I have access to the AutoCAD Drawing Editor when taking the exam? Can I research general knowledge questions I am not familiar with using the AutoCAD software?
A: The proper version of AutoCAD software (AutoCAD Release 14, AutoCAD 2000, AutoCAD LT 2000) must be loaded prior to taking any Assessment exam. Therefore, AutoCAD software is available while working through the exam. The question content is not derived from the facts found in the Help file of AutoCAD, but from practical and functional use of the application itself. Spending the time researching the answer is possible, but may be counterproductive to the completion of the exam.

Q: Can I skip questions or go back and change answers later?
A: Yes.

Q: How will taking an AutoCAD Assessment exam benefit me?
A: Completing an AutoCAD Assessment exam will allow you to assess your current AutoCAD skills. The questions are all grouped into Categories. Once the exam is completed, the electronic test driver generates a diagnostic report. This report will highlight areas in which an individual is either strong or weak in their use of AutoCAD. Since a number of companies conduct their own assessment exams, prior experience in taking the AutoCAD 2000 Assessment exams would enable you to perform well and opens potential doors for employment. If you are already employed, successful completion of the AutoCAD Assessment exams might qualify you for higher pay raises or bonuses.

Q: Who will be interested in my results?
A: Design companies and firms are always looking for the best and most qualified individual. The AutoCAD Assessment exams provide an objective, fact-based way for employers, employment agencies and recruiters to evaluate CAD skills. The AutoCAD Assessment exams could eliminate the need for companies to spend many hours on developing a customized assessment exam. The Assessment Exams could also be used by the company as a marketing tool. This would inform prospective clients that a qualified work force will lead to quality in design. All of this leads to a company that has higher employee morale and is more competitive in the job market.

Q: What resources are available to better prepare for the exams?

A: Your local Autodesk Training Center (ATC) offers AutoCAD 2000 Level I and Level II courses that cover the same technical content you will be expected to know when you take the exams. For the name of the ATC nearest you, call 800-964-6432. You should also see if a local Technical College or University has related courses.

You can prepare for the exam through self-study using Autodesk Official Training Courseware for AutoCAD 2000. The Assessment exams draw directly from the courseware, so we recommend that candidates have a thorough understanding of all the concepts, commands and features covered in the courseware. To order a copy of Autodesk Official Training Courseware for AutoCAD 2000 Level I and Level II, call 800-964-5194. The AutoCAD 2000 Assessment Exam Preparation Manual is also available through Delmar Publishers at (800)998-7498.

License Agreement for Autodesk Press, Thomson Learning™

Educational Software/Data

You the customer, and Autodesk Press incur certain benefits, rights, and obligations to each other when you open this package and use the software/data it contains. BE SURE YOU READ THE LICENSE AGREEMENT CAREFULLY, SINCE BY USING THE SOFTWARE/DATA YOU INDICATE YOU HAVE READ, UNDERSTOOD, AND ACCEPTED THE TERMS OF THIS AGREEMENT.

Your rights:

1. You enjoy a non-exclusive license to use the enclosed software/data on a single microcomputer that is not part of a network or multi-machine system in consideration for payment of the required license fee, (which may be included in the purchase price of an accompanying print component), or receipt of this software/data, and your acceptance of the terms and conditions of this agreement.

2. You own the media on which the software/data is recorded, but you acknowledge that you do not own the software/data recorded on them. You also acknowledge that the software/data is furnished "as is," and contains copyrighted and/or propri-etary and confidential information of Autodesk Press or its licensors.

3. If you do not accept the terms of this license agreement you may return the media within 30 days. However, you may not use the software during this period.

There are limitations on your rights:

1. You may not copy or print the software/data for any reason whatsoever, except to install it on a hard drive on a single microcomputer and to make one archival copy, unless copying or printing is expressly permitted in writing or statements recorded on the diskette(s).

2. You may not revise, translate, convert, disassemble or otherwise reverse engineer the software/data except that you may add to or rearrange any data recorded on the media as part of the normal use of the software/data.

3. You may not sell, license, lease, rent, loan, or other-wise distribute or network the software/data except that you may give the software/data to a student or and instructor for use at school or, temporarily at home.

Should you fail to abide by the Copyright Law of the United States as it applies to this software/data your license to use it will become invalid. You agree to erase or otherwise destroy the software/data immediately after receiving note of Autodesk Press' termination of this agreement for violation of its provisions.

Autodesk Press gives you a **LIMITED WARRANTY** covering the enclosed software/data. The **LIMITED WARRANTY** can be found in this product and/or the instructor's manual that accompanies it.

This license is the entire agreement between you and Autodesk Press interpreted and enforced under New York law.

Limited Warranty

Autodesk Press warrants to the original licensee/purchaser of this copy of microcomputer software/ data and the media on which it is recorded that the media will be free from defects in material and workmanship for ninety (90) days from the date of original purchase. All implied warranties are limited in duration to this ninety (90) day period. **THEREAFTER, ANY IMPLIED WARRANTIES, INCLUDING IMPLIED WARRANTIES OF MERCHANTABILITY AND FITNESS FOR A PARTICULAR PURPOSE ARE EXCLUDED. THIS WARRANTY IS IN LIEU OF ALL OTHER WARRANTIES, WHETHER ORAL OR WRITTEN, EXPRESSED OR IMPLIED.**

If you believe the media is defective, please return it during the ninety day period to the address shown below. A defective diskette will be replaced without charge provided that it has not been subjected to misuse or damage.

This warranty does not extend to the software or information recorded on the media. The software and information are provided "AS IS." Any statements made about the utility of the software or information are not to be considered as express or implied warranties. Autodesk Press will not be liable for incidental or consequential damages of any kind incurred by you, the consumer, or any other user.

Some states do not allow the exclusion or limitation of incidental or consequential damages, or limitations on the duration of implied warranties, so the above limitation or exclusion may not apply to you. This warranty gives you specific legal rights, and you may also have other rights which vary from state to state. Address all correspondence to:

Autodesk Press

3 Columbia Circle

P. O. Box 15015

Albany, NY 12212-5015